Etiopathogenetic Hypotheses of Schizophrenia

Thermodynamic
Structure of
Development

Etiopathogenetic Hypotheses of Schizophrenia

The Impact of Epidemiological, Biochemical and Neuromorphological Studies

Edited by

C.L. Cazzullo, G. Invernizzi, E.Sacchetti and A. Vita

Proceedings of an International Meeting held in Milan, October 3-4, 1986

MTP PRESS LIMITED
a member of the KLUWER ACADEMIC PUBLISHERS GROUP
LANCASTER / BOSTON / THE HAGUE / DORDRECHT

Published in the UK and Europe by
MTP Press Limited
Falcon House
Lancaster, England

British Library Cataloguing in Publication Data

Etiopathogenetic hypotheses of schizophrenia :
  the impact of epidemiological, biochemical and
  neuromorphological studies : proceedings of an
  international meeting held in Milan, October 3-
  4 1986.
  1. Schizophrenia
  I. Cazzullo, C.L.
  616.89´82   RC514

  ISBN-13:978-94-010-7939-6 .   e-ISBN-13: 978-94-009-3207-4
  DOI: 10.1007/978-94-009-3207-4

Published in the USA by
MTP Press
A division of Kluwer Academic Publishers
101 Philip Drive
Norwell, MA 02061, USA

Copyright 1987 MTP Press Limited
Softcover reprint of the hardcover 1st edition 1987

Contents

Preface

Going ahead with their plans, the Association for Research on Schizophrenia (ARS), the Schizophrenia Research Group of the Institute of Psychiatry of the University of Milan and the Tito and Fanny Legrenzi Foundation organized the International Meeting "Etiopathogenetic Hypotheses of Schizophrenia: The Impact of Epidemiological, Biochemical and Neuromorphological Studies", held in Milan on October 3-4 1986.

The Meeting was an excellent occasion for the exchange of information and the sharing of views on the etiological hypotheses of schizophrenia resulting from the latest research in the key areas of epidemiology, biochemistry amd brain imaging.

We are very pleased and proud to have had so many of the leading researchers in schizophrenia with us. Their stimulating and friendly presence and their contributions made this Meeting a great success.

We hope that this book, a collection of the various participants' contributions at the Meeting, will have an equally favourable reception.

The Editors

List of contributors

N.C. Andreasen
The University of Iowa
  Hospitals and Clinics
Department of Psychiatry
500 Newton Road
Iowa City, IA 52242
USA

M. Battaglia
Institute of Psychiatry
Milan University
Policlinico - Pad. Guardia II
via F. Sforza 35
20122 Milan
Italy

L. Bellini
Department of Psychiatry
University of Milan Medical
  School
via G.F. Besta 1
20161 Milano
Italy

L. Bellodi
University of Milan
Institute of Clinical
  Psychiatry
Ospedale S. Paolo
via Di Rudinì 8
20142 Milano
Italy

A. Bocchetta
Department of
  Neurosciences
Clinical Pharmacology
University of Cagliari
via Porcell 4
09100 Cagliari
Italy

F. Brambilla
Psychiatric Hospital "Paolo
  Pini"
via Ippocrate 45
20100 Milano
Italy

A. Breier
Clinical Neuroscience
  Branch, NIHM
Building 10, Room 4N 214
9000 Rockville Pike
Bethesda, MD 20892
USA

A. Calzeroni
Institute of Psychiatry
Milan University
Policlinico - Pad, Guardia II
via F. Sforza 35
20122 Milano
Italy

M. Casacchia
Chair of Clinical Psychiatry
S. Maria di Collemaggio
  Hospital
University of L'Aquila
67100 L'Aquila
Italy

R. Cattaneo
Department of Psychiatry
University of Milan Medical
  School
via G.F. Besta 1
20161 Milano
Italy

C.L. Cazzullo
Institute of Psychiatry
Milan University
Policlinico - Pad. Guardia II
via F. Sforza 35
20122 Milan
Italy

B. Chitkara
Institute of Psychiatry
De Crespigny Park
Denmark Hill
London, SE5 8AF
UK

W. Christie
Klinikum Charlottemburg
Department of Neurology
Spandau Damm 130
D-1000 Berlin 19
West Germany

C. Cignarale
Chair of Clinical Psychiatry
S. Maria di Collemaggio
  Hospital
University of L'Aquila
67100 L'Aquilla
Italy

J.A. Coffmann
Ohio State University
Department of Psychiatry
Room 071, Upham Hall
473 West 12th Avenue
Columbus, Ohio
USA

C. Colombo
Department of Psychiatry
University of Milan Medical
  School
via G.F. Besta 1
20161 Milano
Italy

G. Conte
Institute of Psychiatry
Milan University
Policlinico - Pad. Guardia II
via F. Sforza 35
20122 Milano
Italy

G.U. Corsini
Department of
  Neurosciences
Clinical Pharmacology
University of Cagliari
via Porcell 4
09100 Cagliari
Italy

J.H.W. Crosset
Department of Psychiatry
University of Iowa
500 Newton Road
Iowa City, IA 52242
USA

T.J. Crow
Clinical Research Centre
Division of Psychiatry
Watford Road
Harrow Middx. HA1 3UJ
UK

M. Del Zompo
Department of
    Neurosciences
Clinical Pharmacology
University of Cagliari
via Porcell 4
09100 Cagliari
Italy

A.R. Doran
Clinical Neurosciences
    Branch, NIHM
Building 10, Room 4N 214
9000 Rockville Pike
Bethesda, MD 20892
USA

J.C. Erhardt
Department of Radiology
University of Iowa
500 Newton Road
Iowa City, IA 52242
USA

F. Facchinetti
Psychiatry Hospital "Paolo
    Pini"
via Ippocrate 45
20100 Milano
Italy

C. Faravelli
Institute of Nervous and
    Mental Diseases
Chair of Clinical Psychiatry
Florence University Medical
    School
Policlinico Careggi
via le Morgagni 85
50100 Firenze
Italy

M. Gallucci
Chair of Clinical Psychiatry
S. Maria di Collemaggio
    Hospital
University of L'Aquila
67100 L'Aquilla
Italy

O. Gambini
Department of Psychiatry
University of Milan Medical
    School
via G.F. Besta 1
20161 Milano
Italy

A.R. Genazzani
Psychiatric Hospital "Paolo
    Pini"
via Ippocrate 45
20100 Milano
Italy

R. Goddard
Institute of Nervous and
    Mental Diseases
Chair of Clinical Psychiatry
Florence University Medical
    School
Policlinico Careggi
via le Morgagni 85
50100 Firenze
Italy

W.M. Grove
Department of Psychiatry
University of Iowa
500 Newton Road
Iowa City, IA 52242
USA

C. Guo-Jung
Shanghai Institute of Mental
    Health
Shangai
China

Y. He-Qin
Shanghai Institute of Mental
    Health
Shanghai
China

G. Invernizzi
Institute of Psychiatry
Milan University
Policlinico - Pad. Guardia II
via F. Sforza 35
20122 Milan
Italy

D.P. van Kammen
Veterans Administration
Highland Drive
Pittsburg, PA 15206
USA

W.B. van Kammen
Veterans Administration
Highland Drive
Pittsburgh, PA 15206
USA

S.W. Lewis
Institute of Psychiatry
De Crespigny Park
Denmark Hill
London, SE5 8AF
UK

M. Linnoila
National Institute of
    Alcoholism and Alcohol
    Abuse
DICBR - Laboratory Clinical
    Studies
Bethesda, Maryland
USA

F. Macciardi
University of Milan Medical
    School
Institute of Clinical
    Psychiatry
Ospedale S. Paulo
via Di Rudinì 8
20142 Milano
Italy

G. Martis
Department of
    Neurosciences
Clinical Pharmacology
University of Cagliari
via Porcell 4
09100 Cagliari
Italy

Z. Ming-Dao
Shanghai Institute of Mental
  Health
Shanghai
China

Z. Ming-Yuan
Shanghai Institute of Mental
  Health
Shanghai
China

S. Mulas
Department of
  Neurosciences
Clinical Pharmacology
University of Cagliari
via Porcell 4
09100 Cagliari
Italy

R. Murray
Institute of Psychiatry
De Crespigny Park
Denmark Hill
London SE5 8AF
UK

H.A. Nasrallah
Ohio State University
Department of Psychiatry
Room 071, Upham Hall
473 West 12th Avenue
Columbus, Ohio
USA

T.C. Neylan
Veterans Administration
Highland Drive
Pittsburgh, PA 15206
USA

G. Niu-Fan
Shanghai Institute of Mental
  Health
Shanghai
China

S.C. Olson
Department of Psychiatry
Ohio State University
473 West 12th Avenue
Columbus, Ohio
USA

M.J. Owen
Institute of Psychiatry
De Crespigny Park
Denmark Hill
London, SE5 8AF
UK

S. Pallanti
Institute of Nervous and
  Mental Diseases
Chair of Clinical Psychiatry
Florence Medical School
Policlinico Careggi
via le Morgagni 85
50100 Firenze
Italy

R. Passariello
Chair of Clinical Psychiatry
S. Maria di Collemaggio
  Hospital
University of L'Aquila
67100 L'Aquilla
Italy

J.L. Peters
Veterans Administration
Highland Drive
Pittsburgh, PA 15206
USA

F. Petraglia
Psychiatric Hospital "Paolo
  Pini"
via Ippocrate 45
20100 Milano
Italy

D. Pickar
Cinical Neurosciences
  Branch, NIMH
Building 10, Room 4N 214
9000 Rockville Pike
Bethesda, MD 20892
USA

M. Provenza
University of Milan
Institute of Clinical
  Psychiatry
Ospedale S. Paolo
via Di Rudinì 8
20142 Milano
Italy

L. Pugnetti
Department of Psychiatry
University of Milan Medical
  School
via G.F. Besta 1
20161 Milano
Italy

A.M. Reveley
The Maudsley Hospital
London
UK

M.A. Reveley
The London Hospital (St
  Clement's)
2A Bow Road
London E3 4LL
UK

J. Rosen
Veterans Administration
Highland Drive
Pittsburgh, PA 15206
USA

A. Rossi
Chair of Clinical Psychiatry
S. Maria di Collemaggio
  Hospital
University of L'Aquila
67100 L'Aquila
Italy

E. Sacchetti
Institute of Psychiatry
Milan University
Policlinico - Pad. Guardia II
via F. Sforza 35
20122 Milan
Italy

S. Scarone
Department of Psychiatry
University of Milan Medical
  School
via G.F. Besta 1
20161 Milano
Italy

G. Sedvall
Karolinska Institutet
Department of Psychiatry
Karolinska Hospital
Box 60500, S-104 01,
    Stockholm
Sweden

E. Smeraldi
University of Milan
Institute of Clinical
    Psychiatry
Ospedale S. Paolo
via Di Rudinì 8
20142 Milano
Italy

S. Stocchetti
Institute of Nervous and
    Mental Diseases
Chair of Clinical Psychiatry
Florence University Medical
    School
Policlinico Careggi
via le Morgagni 85
56100 Firenze
Italy

P. Stratta
Chair of Clinical Psychiatry
S. Maria di Collemaggio
    Hospital
University of L'Aquila
67100 L'Aquila
Italy

E. Fuller Torrey
6204 Ridge Drive
Bethesda, MD 20816
USA

G. Vespucci
Chair of Clinical Psychiatry
S. Maria di Collemaggio
    Hospital
University of L'Aquila
67100 L'Aquila
Italy

A. Vita
Institute of Psychiatry
Milan University
Policlinico - Pad. Guardia II
via F. Sforza 35
20122 Milan
Italy

D.R. Weinberger
Section of Clinical
    Neuropsychiatry, NIMH
Saint Elizabeth Hospital
WAW Building
Washington, DC 20032
USA

O.M. Wolkowitz
Clinical Neurosciences
    Branch, NIMH
Building 10, Room 4N 214
9000 Rockville Pike
Bethesda, MD 20892
USA

Z. Zhen-Yi
Shanghai Institute of Mental
    Health
Shanghai
China

EPIDEMIOLOGY

1
Negative symptoms: a follow-up perspective

M. Casacchia, A. Rossi, P. Stratta,
G. Vespucci and C. Cignarale

INTRODUCTION

In reviewing the importance of psychiatric symptoms and signs not only in diagnosis but also in understanding the underlying processess of schizophrenia, Strauss et al. |1| introduced the positive-negative distinction in psychiatry.

Historically, Bleuler considered "negative simptoms" the fundamental deficiency of schizophrenia while Kraepelin viewed the defect/residual state as the consequence of an active psychotic process. Other authors have suggested that negative and positive symptoms represent semi-independent processes |1|.

The study of negative symptoms, generally characterized by a loss of functioning, leads us to consider theoretical and clinical aspects relevant to our knowledge of schizophrenia.

Bilder et al. |2| have suggested that the defect state may not be a monothetic construct, and that within the domain of "type II" schizophrenia, disturbances of thought may be distinguished from those of affect and motivation |3|.

Flat affect, that has long been demonstrated as a central manifestation of "negative" schizophrenia and has been appreciated as prognostically important attribute of this disorder |4|, may be influenced in his clinical evaluation by such components as right hemisphere dysfunction, retardation and extrapyramidal side effects |5|.

Flat affect, emotional withdrawal, motor retardation have been considered to be the key elements of the negative symptom syndrome by most investigators, especially those studying short-term clinical change. In longer term "outcome" perspective non diagnostic variables like work performance and social relationships may be considered "molar variables" while "negative symptoms" may represent "molecular variables".

We have performed a follow-up study to verify if "poor prognosis schizophrenia" as defined by Stephens |6| and Vaillant |7| criteria and measured on prognostic scale by Strauss et al. |8| may predict outcome.

In addiction "negative symptoms" cross sectionally evaluated with Negative Symptoms Rating Scale (NSRS) by Iager et al. |9| were correlated with outcome measure.

METHOD

Subjects: the clinical sample was drawn from a group of admitted patients from a catchment area served by S.M. Collemaggio Hospital in the year 1980–1981. The patients were diagnosed as schizophrenics on the basis of Feighner Criteria. This sample formed the basis for a previous study of short term outcome. 40 male patients (age range 19–35 years) were sampled. During a 5 year follow-up period the patients attended the community service of Psychiatric Hospital; hospital readmission, clinical deterioration, changes of neuroleptic doses were evaluated.

The initial dichotomy was performed on the basis of mixed 13 criteria of Vaillant |7| and Stephens |6| (at least of 7 items present = good prognosis) as reported by Bland et al. |10|. Because the study intended to value the outcome and its clinical correlates and not the prevalence of the outcome itself, 20 patients with good prognostic indicators were matched with 20 patients with poor prognostic indicators. After this "a priori dichotomy", Prognostic Scale |8| was rated for both groups.

The last clinical evaluation was performed during a 10 month period (october 1985 – july 1986). The patients were evaluated on a stable clinical psychopathology state. All patients were treated with neuroleptic drugs.

During the last assessment the patients were rated on Outcome Scale |11|, Brief Psychiatric Rating Scale (BPRS) |12|, NSRS |9|. 26 patients (65.5% of the initial sample) completed the study. Among drop-out patients 5 refused to attend at follow up examination, 4 were treated by another service, 4 were admitted to a psychiatric ward and 1 died.

T test was used for comparison between rating scales. A correlation analysis was performed for outcome score vs. NSRS for both groups of patients.

RESULTS

After the a priori dychotomy the Prognostic Scale was rated; total score was statistically different between groups (good prognosis group score = 33.69 vs. poor prognosis group score = 28.34, t=3.21, d.f. 38, P< 0.01) (Tab. 1). Item n°9 score of Prognostic Scale (flat affect) did not statistically differ between groups.

Table 1. Clinical features in good prognosis group and poor prognosis group

	good prognosis group		poor prognosis group		p
	at baseline				
	pt. 20		pt. 20		
	mean	SD	mean	SD	
Prognostic scale	33.69	5.29	28.34	5.24	.001
Item N°9: affect flattening	3.13	.9	2.47	1.16	NS
	at follow up				
	pt. 13		pt. 13		
	mean	SD	mean	SD	
Outcome Score	7.30	1.43	6.15	1.28	.05
BPRS	53	6.23	50.84	4.89	NS
–anergia factor	10.72	3.6	16.72	4.5	.001
–thinking dist. factor	14.6	3.2	12.4	4.6	NS
NSRS	–9.46	2.93*	–13.30	4.92§	.05

Correlation of outcome vs. NSRS score
* good prognosis group: Pearson r= .68; p<.05
§ poor prognosis group: Pearson r= .00; NS

No statistically significant differences were seen between the two groups on BPRS total score. When Anergia Factor and Thinking Disturbance Factor of BPRS were taken into account separately, only Anergia Factor discriminated the two groups (Tab. 1).

The initial dichotomy was validated on the basis of outcome score (good prognosis group score = 7.30+2.93 vs. poor prognosis group score = 6.15+1.28; p 0.05) and NSRS score (-9.46+2.93 vs. -13.30+4.92; P 0.05).

Moreover when a correlation analysis was performed, Outcome Score in good prognosis group correlated significantly with a low score on NSRS scale (the higher the "outcome score", the better the outcome, the lower the NSRS score) (r=0.68; p 0.05). No correlation was found between Outcome score and NSRS score in the poor prognosis group.

CONCLUSIONS

Negative symptoms have been reported variously embedded with outcome |4|, and in general, psychiatric symptomatology is useful in predicting the profile of recurring psychopathology in any individual patient |13|. Although chronicity may dampen outcome heterogeneity in schizophrenia, it by no means creates differences in negative simptomatology.

Only recently has the development of rating scales specifically designed for negative symptoms allowed us to measure such a symptomatology |14, 15|.

Although methodological and conceptual problems in quantifying and defining negative symptoms remain |16|; nevertheless the use of these instruments may be useful in the process of validation of an hypothetical negative syndrome.

This observation of a significant positive correlation between good outcome score and low NSRS in the good prognosis patient group and the absence of any correlation between the same variables in the poor prognosis group identified by high NSRS score and by a low outcome score indicate that two different patterns of outcome do associate with different "negative" symptoms score. Even if other contributions, such as schizophrenic subtypes |17| can be involved in the development of a poor outcome such an observation deserves further research. McGlashan |18| has recently pointed out that a symptoms analysis in a population of more established and chronically schizophrenic patients, suggested that cross-sectional psychopatology is less restricted in its predictive power than with patients situated earlier in their course of illness.

It would seem more important to verify whether restricted affect (or defect state in general) should be considered an independent predictor of chronicity and/or poor outcome or may serve more as a marker of poor outcome.

In our study flat affect on admission was not defined operationally and this may limit further conclusions.

It is probable "that the inquiry into negative symptoms can provide a fashinating and important pathway into understanding biological/psychological interactions ... genetic, other biological and many psychosocial factors may interact over time in important ways to generate negative symptoms" |19|.

REFERENCES

1. Strauss, jS, Carpenter, WT jr and Bartko, J (1974). The diagnosis and understanding of schizophrenia: Part III. Speculations on the process that underlie schizophrenic symptoms and signs. Schizophrenia Bull, 1 (Experimental Issue N°11), 61

2. Bilder, RM, Mukherjee, S, Rieder, RO et al. (1985). Symptomatic and Neuropsychosocial Components of Defect States. Schizophrenia Bull, 11, N°3, 400

3. Gibbons, RD, Lewine, RRJ, Davis, JM et al. (1985). An Empirical Test of a Kraepelinian vs. a Bleulerian View of Negative Symptoms. Schizophrenia Bull, 11, N°3, 390

4. Carpenter, WT et al. (1978). Signs and symptoms as predictors of outcome: a report from the International Pilot Study of Schizoprenia. Am J Psychiatry, 135, 940

5. Mayer, M, Alpert, M, Stastny, P et al. (1985). Multiple Contributions to Clinical Presentation of Flat Affect in Schizophrenia. Schizophrenia Bull, 11, N°3, 420

6. Stephens, JH, Astrup, C and Mangrum, JC (1966). Prognostic Factors in Recovered and Deteriorated Schizophrenics. Am J Psychiatry, 122, 1116

7. Vaillant, GE (1964). Prospective Prediction of Schizophrenic Remission. Arch Gen Psychiatry, 11, 509

8. Strauss, JS, Carpenter, WT JR (1974). The Prediction of Outcome in Schizophrenia. II Relationship Between Predictor and Outcome Variables. Arch Gen Psychiatry, 31, 37

9. Iager, AC, Kirch, DG and Wyatt, RJ (1985). A Negative Symptoms Rating Scale. Psychiatry Research, 16, 27

10. Bland, RM, Parker, JH and Orn, H (1978). Prognosis in Schizophrenia. Prognostic Predictors and Outcome. Arch Gen Psychiatry, 35, 72

11. Strauss, jS and Carpenter, WT Jr (1972). The Prediction

of Outcome in Schizophrenia. Caratheristic of Outcome I. Arch Gen Psychiatry, 27, 739

12. Overall, GE and Gorham, DR (1962). The brief psychiatric rating scale. Psychological Reports, 10, 799

13. Strauss, JS and Carpenter, WT (1978). The prognosis of schizophrenia: Rationale for a multidimensional concept. Schizophrenia Bull, 4, 56

14. Andreasen, NC (1982). Negative symptoms in schizophrenia: Definition and reliability. Arch Gen Psychiatry, 39, 784

15. Andreasen, NC and Olsen, S (1982). Negative vs. positive schizophrenia: Definition and validation. Arch Gen Psychiatry, 39, 789

16. Sommers, AA (1985). "Negative Symptoms": Conceptual and Methodological Problems. Schizophrenia Bull, 11, N°3, 364

17. Kendler, KS, Gruenberg, AM and Tsuang, MT (1984). Outcome of Schizophrenics Subtypes Defined by Four Diagnostic Systems. Arch Gen Psychiatry, 41, 149

18. McGlashan, TH (1986). The Prediction of Outcome in Chronic Schizophrenia. IV. The Chestnut Lodge Follow-up Study. Arch Gen Psychiatry, 43, 167

19. Strauss, jS (1985). Negative Symptoms: Future Developments of the Concept. Schizophrenia Bull, 11, N°3, 457

2
Obstetric complications and cerebral abnormalities in schizophrenia
M.J. Owen, S.W. Lewis and R.M. Murray

INTRODUCTION

The idea that adverse circumstances at or around the time of birth
might play a part in the aetiology of neuropsychiatric disorders
owes much to a series of studies by Pasamanick and his colleagues
in the 1950's. They assessed birth certification retrospectively
in several large samples of children and compared their documented
obstetric histories with those of normal controls. These studies
established that rates of recorded obstetric complications (OCs)
were increased in children with cerebral palsy, epilepsy, mental
retardation, behaviour disorders, reading disabilities and tics
(19). This prompted a reformulation of ideas concerning the pathogenesis of several of these conditions. For example, the aetiology
of mental retardation, once considered almost entirely hereditary
or familial, was increasingly regarded as multifactorial with
environmental factors as important as genetic. Pasamanick (19)
postulated "a continuum of reproductive casualty, consisting of
brain damage incurred during the prenatal and perinatal periods as
a result of abnormalities during these periods, leading to a
gradient of injury extending from foetal and neonatal death, through
cerebral palsy, epilepsy, behaviour disorder and mental retardation".
    Pasamanick confined his inquiries to childhood disorders
although he noted certain epidemiological similarities, such as
the seasonality of birth effect, between some of these conditions
and schizophrenia. However, twenty years earlier Aaron Rosanoff
and his colleagues (28) had investigated early antecedents of
schizophrenia as part of a study of 1014 pairs of mentally
disordered twins. Among those monozygotic pairs discordant for
schizophrenia these authors claimed to be able to isolate a group
of cases in which the illness resulted from a "partial decere-
bration" due to birth injury. Rosanoff and his colleagues
distinguished between different classes of effect of cerebral
injury at birth: "immediate effects", such as stupor, "permanent
effects" such as mental deficiency, and "sequels, appearing after
an interval of weeks, months or years following the trauma, such
as epilepsy or psychotic disease.....". Moreover, post-mortem
evidence suggested to Rosanoff et al that cerebral atrophy,

particularly frontolateral, characterised the brains of a proportion of schizophrenic patients, and that these cases represented
"the remote results of a cerebral birth trauma". Rosanoff capably
explored the predictions of his hypothesis, which included
particularly the relatively asymptomatic nature of the original
lesion, a younger age at onset in cases so affected, and a predominance in first born males. Both he and Pasamanick conceived of a
reduced familial predisposition in cases with a history of OCs in
comparison to cases with no such history.

In the 1960's two groups undertook retrospective studies of OCs
in cohorts of schizophrenic patients. Lane and Albee (7) found that
the mean birthweight, as recorded on birth certification, of 52
schizophrenic adults was significantly less than that of their 115
siblings. Pollack's group reported that nine schizophrenic patients
whose mothers reported "severe paranatal complications" were
younger at first hospitalisation and showed a poorer outcome than
43 patients without such backgrounds (22,23). In a second series of
32 schizophrenic patients the same group reported a higher OC score
compared with healthy siblings, although no significant difference
in birthweight was seen (36,37).

McNeil and Kaij (13) reviewed these findings and added data
of their own showing OCs to be increased in the blindly-rated
birth records of 54 process schizophrenic patients compared to 100
contemporary births matched for sex, maternal age, parity and social
class. The authors went on to re-examine existing studies of
identical twins discordant for schizophrenia finding that, in 72%
of 39 existing pairs, the affected (or more severely affected) twin
had suffered the more birth complications. Jacobsen and Kinney (4)
studied midwife records and noted that prolonged labour best
distinguished 63 people with schizophrenia from 53 controls;
prematurity was nearly twice as common in the former although this
difference failed to reach significance. Parnas et al (18) divided
the children of schizophrenic mothers into those who developed
schizophrenia, borderline schizophrenia, and normal individuals;
those with schizophrenia had the most, and the borderlines the least
complicated births. It should be noted that, because these three
Scandinavian studies examined data recorded at birth, reporting
bias cannot explain the results.

Obstetric Complications and Adult CT Scan Findings

The idea that OCs may be risk factors for schizophrenia has been
given further impetus by studies using computerised tomography. The
finding of enlarged cerebral ventricles in approximately one third
of adults with schizophrenia has been frequently replicated (16).
The weight of evidence suggest that increased ventricular size is
not progressive and is present at the onset of, and probably
antedates, the illness itself. Links between CT scan findings and
early neurological insults such as OCs have been found in four
studies from the Institute of Psychiatry and its related hospitals,
as well as in three from elsewhere.

Reveley et al (25) reported that OCs were associated with
larger ventricles in a sample of normal twins. Moreover, a history
of OCs was obtained only from those schizophrenic twins who did not

10

have a family history of major psychiatric disorder; ventricular size in this group was twice that of schizophrenic twins with such a family history. Roberts (26) noted that birth and early environmental hazards was associated with later CT scan abnormalities among general psychiatric patients, while Turner et al (33) found that lateral ventricular size was correlated with OCs among 50 first episode schizophrenic patients.

Pearlson et al (21) reported that, of 19 schizophrenic patients, the four with a history of early developmental insult showed a trend towards larger ventricles. Schulsinger et al (29) found that birthweight predicted ventricular-brain ratio (VBR) among 29 offspring of schizophrenic mothers, and DeLisi et al (1) demonstrated that a history of OCs was predictive of VBR among 20 schizophrenics and their 10 well siblings.

In a pilot survey of obstetric history among 955 psychiatric patients Lewis and Murray (9) blindly rated definite OCs as present in 17% of the 207 patients with schizophrenia as against 8% of the 748 with other diagnoses. OCs were significantly more common in patients with schizophrenia than in the sample as a whole ($X^2 = 13.9$; $p < 0.001$) and as compared with the 203 neurotic subjects ($X^2 = 12.8$; $p < 0.001$). 236 individuals had undergone a CT scan. Radiologists reports were abnormal in 42% of those with a definite OC compared with 20% of those with an equivocal OC and only 13% of those with no known OC ($X^2 = 12.5$; $p < 0.001$).

Several groups (e.g. 20,28) have argued that if OCs are of aetiological significance in later psychiatric disorder, then they ought to be more common in those without any manifest genetic predisposition. McNeil (12) was unable to confirm this, but in the study just discussed we found a history of OCs in 6% of schizophrenics with a history of psychiatric admission in a first degree relative (i.e. the same proportion as among neurotic patients) but in 24% of non-familial schizophrenics.

PRESENT STUDY

In view of the trends suggested by the existing literature we decided to examine certain features of the illness in a cohort of patients with schizophrenia in whom obstetric complications and family history had been rated blindly. Specifically we were interested in answering the following questions; 1) What is the relationship between OCs and manifest genetic predisposition? If Rosanoff's original conception of a reduced family history in cases where OCs occurred is correct, then this would provide compelling evidence that OCs or some factor associated with them are of aetiological importance in schizophrenia. 2) What is the relationship between OCs and the neuroradiological abnormalities found in schizophrenia? We predicted that, if OCs are associated with early brain damage, then CT stigmata should be more common in patients who had a difficult birth. The two CT abnormalities most consistently noted in schizophrenia are ventricular enlargement and cortical surface atrophy. These do not appear to be correlated in the schizophrenic population as a whole (15,35) yet there is evidence that both may be associated with an absence of family history (10,17). The present study therefore set out to examine the relationship to obstetric history of both ventricular size and cortical surface atrophy. 3) What is the relationship between familial predisposition and CT abnormalities in schizophrenia? If OCs are associated with neuroradiological abnormalities then the latter ought to be more common among non-familial schizophrenics. The weight of evidence suggests that an inverse relationship does exist between the presence of cerebral pathology and manifest genetic predisposition (10). If this is indeed the case then it would suggest that whatever events underlie the neuroradiological abnormalities are of aetiological importance (16).

Method

All patients aged 16-50 years who had been discharged with a diagnosis of schizophrenia or schozoaffective psychosis following their first admission to the Maudsley Hospital between January 1981 and December 1984 were assessed for inclusion. 149 such patients were identified. The case records of each patient were then examined and 39 subjects (26%) excluded because of inadequate information concerning obstetric history. The remaining 110 individuals (67 males and 43 females) served as subjects.
   The following data were obtained concerning the index admission of each patients. i) Obstetric Complications: Information about gestation and birth was abstracted from each case summary and rated by RMM who was blind to the patients' identity and any details of the case, including family history. He was also blind to diagnosis because the ratings were made as part of a study comparing the prevalence of OCs in different diagnostic groups (9). Each patient was assessed according to the criteria in Table 1 and rated as having a history of either definite, equivocal or no OCs. 2) Early Cerebral Damage (ECD): At the same time any information relating to possible brain damage when aged 10 or under was assessed. Patients were rated as positive (ECD+) if a definite history of encephalitis, head injury leading to hospital admission or named congenital cerebral disorder had been noted. 3) Family History: This was rated on a separate occasion by RMM who was again blind to the patients' identity and any details of the case including obstetric history. Patients were adjudged to be "family history positive" (FH+) if there was a history in any first degree relative of psychiatric illness requiring admission to hospital. Family history and data concerning early development of patients in the Maudsley Hospital is usually ascertained from interview with at least one relevant relative in addition to the patient.
   CT scans were performed on the Maudsley Hospital 1010 scanner. Measurements of lateral ventricular area and size of the intracranial space were made using mechanical planimetry from the cut showing the largest area of lateral ventricle. Each scan was measured independently by MO and SWL blind to the patients' identity. Inter-rater reliability was good (r = 0.98). VBRs were calculated from the mean values of both investigators according to the method of Synek and Reuben (32).

Table 1 Obstetric Complications Scale

Definite	Equivocal
Antepartum	
1. Rubella or syphilis	
2. Phesus incompatability	
3. Pre-eclampsia: severe and/or leading to early induction of hospitalisation	Pre-eclampsia not otherwise specified (NOS)
4. APH or threatened abnortion	
Intrapartum	
5. Premature rupture of membranes, 24 hours	
6. Labour $>$ 36 hours or $<$ 3 hours	Labour $>$ 24 hours or "long/difficult/precipitate" NOS
7. Twin birth, complicated	Twin birth NOS
8. Cord prolapse	Cord knotted or round neck
9. Gestational age $<$ 37 weeks or $>$ 42 weeks	"Premature" or "postmature" NOS
10. Caesarian, complicated or emergency	
11. Breech or abnormal presentation	
12. High or "difficult" forceps	Forceps or other instrumental delivery, NOS
13. Birthweight $<$ 4½lbs (2000g)	$<$ 5½lbs (2500g) or "small" NOS
Postpartum	
14. Incubator $>$ 4 weeks	Incubator/resuscitation/"blue" NOS
15. -	Gross physical anomaly

This scale represents a consensus derived from 6 scales used previously, three from the obstetric and three from the psychiatric literature (refs. available from authors).

Atrophy of the cortical surface was also assessed. Three indices were rated following Weinberger et al (35) and Ron (27); 1) cortical sulcal widening, 2) widening of Sylvian structures and 3) widening of the interhemispheric fissure. Sulcal widening was rated on a 4 point scale (0-3). The standards of Nasrallah et al (14) were employed to define the lower limits of each category which corresponded to those employed by Weinberger et al (35). The maximum width of the Sylvian structures (either fissure or cistern) was rated on a 4 point scale (0-3) and that of the interhemispheric fissure on a 3 point scale (0-2). Reference was made to photographs of standard scans defining the lower limit of each category which again corresponded approximately to those of Weinberger et al (35). The scans were assessed independently by the two raters blind to the patients' identity was well as the numerical values of the VBRs. Where the raters disagreed (only 13% of ratings) the scans were re-examined independently. If disagreement persisted then a consensus was obtained. Usually this was the lower of the two discrepant ratings.

Family History and Obstetric Complications

Of the 110 patients, 18 (16%) were rated as having a definite
history of OCs (15 male) and 13 (12%) a history of equivocal OCs (10
male). Five patients had a history of ECD (3 male). 23 (21%)
patients were rated as FH+ (17 male); none of these had a history of
definite OCs or a history of ECD. There was therefore a marked
inverse relationship between the presence of definite OCs and aFH
of mental illness ($z = 2.06$, $p < 0.05$, 2 tailed). Although there was
no overlap between the definite OC group and the FH+ group two
patients who were rated as FH+ had a history of equivocal OCs. In
addition, one FH+ patient with a history of definite OCs was
excluded from the study by an investigator blind to the OC ratings
when he was discovered to have been first admitted to the Maudsley
Hospital before January 1981.

CT Findings

69 patients were scanned, and data are available concerning ventricular size for 65 and cortical surface atrophy for 66. These
interim data are summarised in Fig. 1. Of the patients scanned, 15
had a history of definite OCs (13 male) and in 5 the history was
equivocal (4 male). Four patients had significant ECD (2 male) and
11 were rated as FH+ (10 male). There were no significant differences
between the ages of the subjects in the various groups (mean age
of whole sample = 28 years).
    Overall there was a trend for patients with a history of
definite OCs to have larger VBRs (definite OC group mean $\pm$ SD =
8.35 $\pm$ 3.59, patients without history suggestive of OCs or ECD mean
$\pm$ SD = 6.94 $\pm$ 2.84), but this was not statistically significant
($t = 1.48$, df 54). In contrast, patients with a positive family
history had significantly smaller VBRs than those without (means =
5.08 vs 7.82, $t = 2.82$, df 62, $p < 0.02$, 2 tailed).
    The sum of the three ratings of cortical atrophy was calcul-
ated to produce a "cortical score" for each patient (27). Nine of
the 15 patients with a history of definite OCs (60%) had a cortical
score of greater than zero compared with 19 out of the 46 (41%)
without a history of OCs. This difference was not statistically
significant ($z = 0.96$). Only 2 of the 11 patients with a family
history (18%) had a cortical score of one or over compared with
27 of the 55 patients (49%) without. This trend approached
statistical significance ($z = 1.54$).
    Fig. 1 demonstrates that, among patients with a history of
definite OCs, there appeared to be a correlation between large VBRs
and cortical atrophy whereas no such relationship existed in the
remaining subjects. Moreover, relatively more patients with definite
OCs appeared to have both large VBRs and cortical scores of one or
over. Indeed 9 out of 15 patients with definite OCs had both a VBR
of greater than the grand mean (8.25) and a cortical score of one
or over, compared with 7 out of 46 patients without an abnormal
obstetric history. This difference was highly statistically
significant ($z = 3.06$, $p < 0.002$, 2 tailed). The FH+ group showed
the opposite tendency; 8 out of 11 had both a VBR of less than the
grand mean and a cortical score of zero compared with 20 out of the

14

55 patients without a FH (z = 1.88, p$<$0.05, 1 tailed).

FIGURE 1

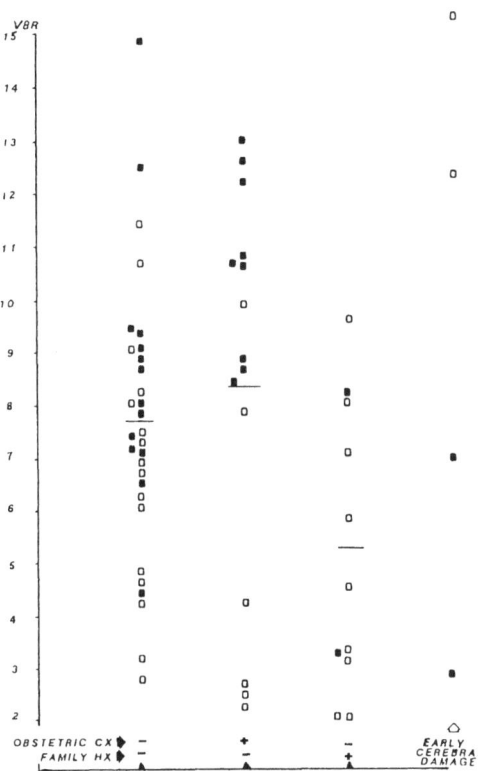

Figure 1 shows the CT findings for the 4 groups: a) neither OCs or FH+, 6) OCs but FH-, c) FH+ but no OCs, d) early cerebral damage. Filled squares - cortical score of more than one, open squares - cortical score of zero.

Discussion

This study shows an inverse relationship between a family history of major psychiatric illness and a history of obstetric complications among schizophrenics, and confirms previous evidence that familial cases are less likely than their non-familial counterparts to show abnormalities on the CT scan. On the other hand, schizophrenics who have suffered hazardous births are especially likely to possess such stigmata. These results provide further evidence that adverse circumstances at, or around, the time of birth leading to brain damage are of aetiological importance in some cases of schizophrenia. Of course, it is logically possible that both OCs and brain damage are secondary to a third factor,

15

such as intrauterine viral infection, but this is a less parsim-
onious explanation.

Our results suggest that while VBR and atrophy of the cortical
surface are not correlated in the majority of schizophrenics, the
brain damage associated with OCs is reflected in both enlargement
of the cerebral ventricles and in surface atrophy. This implies
that there is heterogeneity of cerebral pathology in schizophrenia,
presumably reflecting the fact that OCs are not the only means by
which brain damage of aetiological significance can occur. Given
the relative insensitivity of CT scanning, it is unclear whether,
if schizophrenia is to ensue, damage must occur to certain specific
neuronal structures or whether the probability of the illness
developing is related predominantly to the amount of tissue damaged.

If we accept that brain damage associated with obstetric
hazard is of aetiological importance in schizophrenia, then two
questions immediately arise. First, how is the infant brain
damaged? Second, how does damage at birth result in adult
psychiatric disorder?

Obstetric Hazard and the Infant Brain

The evidence as it stands points to the importance of cerebrovascular events in the production of brain damage at birth. The
recent introduction of ultrasound brain scanning has shown that
intraventricular and periventricular haemorrhage are much more
common than hitherto supposed, occurring, for example, in approximately 50% of preterm babies of 1500gm or less (5,6). The exact
pathogenic mechanism is still controversial, but both hypo- and
hyperperfusion of cerebral blood vessels have been implicated. Ante-
partum haemorrhage, birth asphyxia, recurrent apnoea and septicaemia
are common antecedents (30). The initial bleeding appears to arise
from the fragile capillaries of the subependymal plate. It may
remain confined to the germinal layer, spread into the ventricles,
or extend into the parenchyma. Periventricular infarction may also
occur and periventricular leukomalacia and cysts may develop. The
bleeding is probably of little consequence in the absence of either
ventricular dilatation or parenchymal extension. Even when
ventricular dilatation does occur it is usually transient, but in
those in whom it persists it may reflect partial obstruction of CSF
by blood clot, or it may be secondary to minor loss of cerebral
tissue through hypoxic-ischaemic damage.

The most severely affected infants die, while others develop
cerebral palsy or mental retardation. Recently several groups have
examined the implications of more subtle lesions for subsequent
development. Leichty et al (8) followed 81 infants who underwent CT
scanning, and at one year showed an association between lower
developmental scores and larger ventricles. As the size of the
haemorrhage increases so the proportion with neurodevelopmental
abnormality increases; the worst outcome is in those with
parenchymal extension. McCarton-Daum et al (11) noted that evidence
of periventricular leucomalacia on CT scan at 40 weeks was a strong
prediction of deviant development at 18 months, while Fletcher et
al (3) reported that increased and progressive ventricular dilatation

predicted subsequent deficits on IQ and performance scores.

The most systematic study is that of Stewart et al (31) who followed up 342 premature infants to one year. Twenty six infants with grossly enlarged ventricles or obvious cerebral atrophy had a very poor neurological outlook while the 275 with normal scans or uncomplicated intraventricular haemorrhage almost invariably developed normally. Among the intermediate group with ventricular dilatation, minor disorders particularly abnormalities of tone and reflexes were common. Preliminary analyses suggest that both ventricular enlargement and the predictive power of the initial findings at brain scan persist for at least 4 years (Stewart, personal communication).

Older children with a history of OCs, but who were born before the availability of brain scanning, also have abnormalities of tone and reflexes as well as developmental and cognitive impairments and an excess of psychiatric disorder (2). Moreover, such 'soft' neurological signs occur more commonly in schizophrenics than in other psychiatric patients and normal controls (e.g. 38). Quitkin and his colleagues (24) have reported that 'soft' signs are correlated with intellectual deficits and claim that there is a subgroup of young schizophrenics who have premorbid asociality, academic difficulty, evidence of 'organicity' on psychological testing, poor prognosis, and poor response to phenothiazines. These authors noted that such patients were characterised by lack of family history and considered that OCs might be an aetiological factor.

Early Brain Damage and Adult Psychiatric Disorder

Clearly then, the results of the above studies support Pasamanick's conception of a "continuum of reproductive casualty" with the implication of cerebrovascular events at or around the time of birth. Nevertheless, the vast majority of individuals who undergo difficult births suffer no psychiatric disorder in later life, and only a minority of those with persistent ventricular dilatation ever have any psychiatric sequelae. As we have already indicated the location and/or the extent of the lesion will be important in determining the likelihood of subsequent psychiatric disorder. This may also depend upon a complex interplay between brain damage and influences including neuronal maturation, individual genetic differences and the extent to which interaction and relationships with important figures such as parents, teachers and employers are impaired.

But why should brain damage in early life not manifest itself in psychosis until young adulthood? A plausible suggestion is that this reflects an interaction between the lesion and maturation of the brain (stated recently in 34). According to this hypothesis, psychotic symptoms have a characteristic age of onset, not because the critical lesion occurs then, but because the brain systems it involves only become functionally active at that time. This is not necessarily to say that the affected individual will show no abnormality in childhood. It is of interest that preliminary analysis of our data showed a particular excess of schizoid traits in the premorbid personalities of those schizophrenic patients with a history of OCs. Moreover, as we have seen there is evidence that a proportion of such patients show neurological and/or intellectual impairment.

In conclusion, our study adds to the growing body of evidence suggesting that OCs are of aetiological significance in some cases of schizophrenia. We have also provided evidence that cerebral abnormalities are specially common in these patients. Further study of such individuals might provide considerable insight into the pathogenesis of schizophrenia.

REFERENCES

1 DeLisi, LE, Goldin, LR, Hamovit, JR, Maxwell, ME, Kurtz, D and Gershon, ES (1986) A family study of the association of increased ventricular size with schizophrenia. Arch Gen Psychiat, 43, 148-153.

2. Dunn, HG, Crichton, JW, Robertson, AM, McBurney, K, Grunau, RVE and Penfold, PS (1985) Evolution of minimal brain dysfunctions to the age of 12 to 15 years. In:Dunn, HG (ed) The Sequelae of Low Birth Weight. The Vancourver Study. (Blackwell, Oxford).

3. Fletcher, JM, Levin, HS and Landry, SH (1984) Behavioural consequences of cerebral insult in infancy. In: Almli, CR and Finger, S (eds) Early Brain Damage, Vol. 1. (Academic Press, New York and London).

4. Jacobsen, B and Kinney, DK (1980) Perinatal complications in adopted and non-adopted schizophrenics and their controls: Preliminary results. Acta Psychiat Scand, 62 (suppl. 285), 337-346.

5. Lancet Editorial (1984) Ischaemia and haemorrhage in the premature brain. Lancet ii, 847-848.

6. Lancet Editorial (1985) Post-haemorrhagic ventricular dilatation in infants. Lancet ii, 1280-1282.

7. Lane, EA and Albee, GW (1966) Comparative birth weights of schizophrenics and their siblings. J Psychol, 64, 227.

8. Leichty, EA, Gilmore, RL, Bryson, CQ and Bull, MJ (1983) Outcome of high-risk neonates with ventriculomegaly. Devel Med Child Neurol, 25, 162-168.

9. Lewis, SW and Murray,RM (1987) Obstetric complications, neurodevelopmental deviance and risk of schizophrenia. Presented at the International Symposium on Genetic Research in Psychiatry, Berlin, Sept. 1986. (Proceedings in press).

10. Lewis, SW, Reveley, AM, Reveley, MA, Chitkara, B and Murray, RM (1987) The familial-sporadic distinction as a strategy in schizophrenia research. Brit J Psychiat,(in press).

11. McCarton-Daum, C, Danziger, A, Ruff, H and Vaughan, HG (1983) Periventricular low density as a predictor of neurobehavioural outcome in very low birth weight infants. Devel Med Child Neurol, 25, 559-565.

12. McNeil, TF (1986) Perinatal factors in the development of schizophrenia. Paper presented at Workshop on Biological Perspectives in Schizophrenia. Dahlem, Berlin.

13. McNeil, TF and Kaij, L (1978) Obstetric factors in the development of schizophrenia: Complications in the births of preschizophrenics and in reproduction by schizophrenic parents. In: Wynne, LC, Cromwell, RL and Matthysse, S (eds) pp. 401-429 (Wiley,

New York).

14. Nasrallah, HA, Kuperman, S, Jacoby, CG, McCalley-Whitters, M, Hamra, B (1983) Clinical correlates of sulcal widening in chronic schizophrenia. Psychiat Res, 10, 237-242.

15. Nasrallah, HA, McCalley-Whitters, M. Jacoby, CG (1982) Cortical atrophy in schizophrenia and mania: A comparactive CT study. J Clin Psychiat, 43, 439-441.

16. Owen, MJ and Lewis, SW (1986) Lateral ventricular size in schizophrenia. Lancet ii, 223-224.

17. Oxiensterna, G, Bergstrand, G, Bjerkenstedt, L, Sedvall, G and Wik, G (1984) Evidence of disturbed CSF circulation and brain atrophy in cases of schizophrenic psychosis. Brit J Psychiat, 144, 654-661.

18. Parnas, J, Schulsinger, F, Teasdale, TW, Schulsinger, H, Feldman, PM and Mednick, SA (1982) Perinatal complications and clinical outcome within the schizophrenia spectrum. Brit J Psychiat, 140, 416-420.

19. Pasamaick, B (1961) Epidemiologic investigations of some prenatal factors in the production of neuropsychiatric disorder. In: Hoch, PH and Zubin, J (eds) Comparative Epidemiology of the Mental Disorders (Grune & Stratton, London).

20. Pasamanick, B, Rogers, ME and Lilienfeld, A (1956) Pregnancy experience and the development of childhood behaviour disorder. Am J Psychiat, 112, 613-618.

21. Pearlson, GD, Garbacz, DJ, Moberg, PJ, Ahn, HD and DE Paulo, JR (1985) Symptomatic familial, perinatal and social correlates of computerised axial tomography (CAT) changes in schizophrenics and bipolars. J Nerv Ment Dis, 173, 42-50.

22. Pollack, M and Greenberg, IM (1966) Paranatal complications in hospitalised schizophrenic and nonschizophrenic patients. J Hillside Hosp, 15, 191.

23. Pollack, M, Levenstein, S and Klein, DF (1968) A three year post-hospital follow up of adolescent and adult schizophrenics. Am J Orthopsychiat, 38, 94.

24. Quitkin, F, Rifkin, A and Klein, DF (1976) Neurological soft signs in schizophrenia and character disorders. Arch Gen Psychiat, 33, 845-853.

25. Reveley, AM, Reveley,MA and Murray, RM (1984) Cerebral ventricular enlargement in non-genetic schizophrenia: A controlled twin study. Brit J Psychiat, 144, 89-93.

26. Roberts, J (1980) The Use of the CT Scanner in Psychiatry. M.Phil. Thesis, University of London.

27. Ron, MA (1983) The alcoholic brain: CT scan and psychological findings. Psychol Med Monograph Suppl 3, (Cambridge University Press).

28. Rosanoff, AJ, Handy, LM, Rosanoff-Plasset, I and Brush, S (1934) The etiology of so-called schizophrenic psychoses. Amer J Psychiat, 91, 247-286.

29. Schulsinger, F, Parnas, J, Petersen, ET et al (1984) Cerebral ventricular size in the offspring of schizophrenic mothers: A preliminary study.Arch Gen Psychiat, 41, 602-606.

30. Sinha, SK, Davies, JM, Sims, DG and Chiswick, ML (1985) Relation between periventricular haemorrhage and ischaemic brain lesions diagnosed by ultrasound in very preterm infants. Lancet

11, 1154-1156.

31. Stewart, AL, Renolds, EOR. Hope, PL et al (1986) Probability of neurodevelopmental disorders estimated from ultrasound appearance of brain in very preterm infants. (in press).

32. Synek, V, and Reuben, JR (1976) The ventricular brain ratio using planimetric measurement of EMI scans. Brit J Radiol, 49, 233-237.

33. Turner, SW, Toone, BK and Brett-Jones, JR (1986) Computed tomographic scan changes in early schizophrenia - preliminary findings. Psychol Med, 16, 219-226.

34. Weinberger, DR (1986) The pathogenesis of schizophrenia: A neurodevelopmental theory: In: Nasrallah, HA and Weinberger, DR (eds) Handbook of Schizophrenia, Vol. 1. The Neurology of Schizophrenia. (Elsevier).

35. Weinberger, DR, Torry, EF, Neophytides, AN and Wyatt, RJ (1979) Structural abnormalities in the cerebral cortex of chronic schizophrenic patients. Arch Gen Psychiat, 36, 935-939.

36. Woerner, MG, Pollack, M and Klein, DF (1971) Birth weight and length in schizophrenics, personality disorders and their siblings. Brit J Psychiat, 118, 461.

37. Woerner, MG, Pollack, M and Klein, DF (1973) Pregnancy and birth complications in psychiatric patients: A comparison of schizophrenic and personality disorder patients with their siblings. Acta Psychiat Scand, 49, 712.

38. Woods, BT, Kinney, DK and Yurgelun-Todd, D (1986) Neurologic abnormalities in schizophrenic patients and their families. Arch Gen Psychiat. 43, 657-663.

3
Genetic notes of schizophrenic disorders

F. Macciardi, M. Provenza,
L. Bellodi and E. Smeraldi

Few psychiatric research fields are so widespread as that of familial studies of schizophrenic disorders. Many observations have been produced in this field but it is very difficult to weigh correctly all the different factors that affect any quantification of these phenomena. In any case, the presence of other psychiatric disorders similar to dissociative manifestations in the families of schizophrenic probands has always been so evident that they have always been included in the definition of the disease, starting with Kraepelin.

The degree of this aggregation varies according to how narrowly the diseases is defined. Even according to the DSM III categories, which are more reliable than mere clinical concepts, the risk for schizophrenic disorders in probands relatives is higher than in the control population.

Whereas general population risk is valued to be 1 or 2% it is increased from 5 to 10 times in relatives as Table 1 shows.

Table 1.   Family Studies on Schizophrenia.

	PARENTS	SIBLINGS	CHILDREN
Pooled European Populations (1)	5.6%	10.1%	12.8%
Italian Population (2)	4.3%	7.9%	
American Population (1)	from 5.5% to 8.1%		

Such a risk is not very high for inherited disease (1), since, for example, the relatives of probands with Affective Disorders

have an estimated risk around 20% to be affected by Affective
Disorders (2). However, this risk seems to be higher than the risk
values for other organic diseases, as Diabetes and Hypertension,
for which genetic susceptibility has been demonstrated or is widely
accepted.

The most important aspect of these risks is the constantly
observed effect on them of the type of relationship, that is, there
is a higher values for siblings than for parents. This effect is
essential to any interpretation and may be the result of many
different causes (Table 2).

Table 2. Hypothethical causes of different risk values in
parents and in siblings of Schizophrenics.

---

1) Generational effect
2) Dominant effect
3) Variation in penetrance
4) Assortative mating
5) Differential fertility

Some of these are genetic in nature, such as the dominance
effect. Others, which are not genetic in origin, such as
assortative mating and generational effects, are revealed whenever
an estimate of genetic parameters is attempted (3), a typical
example of the structural difficulties of genetic studies when
applied to Schizophrenic Disorders.

The effect of the type of relationship on the observed risk for
Schizophrenic Spectrum Disorders enabled us to observe something
else in our sample. In fact, if we do not take into account the
risk for the Schizophrenic Spectrum Disorders but the risk for
Affective Disorders, we confront two antithetical theses (4,5,6).
Some studies seem to indicate that susceptibility for Schizophrenic
Disorders protects from Affective Disorders, so that relatives of
schizophrenic patients may show a very low risk for Affective
Disorders. On the other hand, in other studies an 8% risk for
Affective Disorders has been observed in relatives of schizophrenic
patients. In our current data, these theses intersect with the
effects of the type of relationship, because the risk for Affective
Disorders is low, from 1% to 2%, for siblings, and higher, about
5%, for parents. This pattern is exactly opposite to that for the
risk for Schizophrenic Spectrum Disorders. Even if we do not
clearly understand the meaning of this finding, we think that this
recent observation further complicates the studies of familial
risk. Might this risk be additive to the specific risk for

Schizophrenic Spectrum Disorders? Why is a schizo-affective condition not set up even if both liabilities are present in the same family? Does this finding affect only the prognosis of the illness in probands and not the presence of the illness in relatives?

Crow (13) raised the question of how to explain with confidence two well-studied findings in the familial studies of Affective Disorders, i.e., the higher prevalence of homotypic disorders in probands' siblings than in parents, with a progressive increase of Affective Disorders in successive birth-cohorts and earlier age at onset in each twentieth-century birth-cohort. The age at onset anticipation is essential for Crow's theory, since it is assumed to be an index of increasing genetic severity. This pure phenomenon does not imply per se a genetic background: it could be the result of alteration of non-genetic factors from generation to generation which involve an easier phenotypical manifestation of the affected genotype, with consequent higher prevalence of Affective Disorders and hence different breakdown. However, we observed that there is earlier age at onset in the offspring of parents with Affective Disorders, which overcomes the mean trend among generations (14). This is a specific characteristic of those subjects with at least one parent affected whatever the birth-cohort of the subject, all other factors being equal. We are confident that we are dealing with a direct involvement of the genetic susceptibility system in the disease. In these well-identifiable families the penetrance parameter seems to change from generation to generation. We do not yet know the reason for this increment of penetrance, but the existence of affected parents could act in earlier breakdown as a non genetic-stressor, since the onset in the parents almost always occurs before than in the offspring. Crow suggests that there is a genetic continuum in susceptibility system for Affective and Schizophrenic Disorders, characterized by increasing severity, and supports his theory with the observation that genetic mechanisms with similar effects seem to be involved in both disorders.

Certainly there is a greater risk for Schizophrenic Disorders in siblings than in parents, since all the tested populations show similar figures for the trend towards higher prevalence. The anticipation of the age at onset is more intriguing in schizophrenic patients because we can hypothesize that they have a reduced fertility. We could not demonstrate any specific effect of parental status on the offsprings breakdown (7), while other variables, such as clinical subtype in the proband and the sex of the relative, accounted for the variability of onset, even when the number of sibs in each family was included as an experimental index of reduced fertility. Due to these differences, although we acknowledge the heuristic value of Crow's hypotesis, we cannot accept it without further specific demonstration. The psychiatric disorders in relatives of affective and schizophrenic probands are not limited to Affective or Schizophrenia Disorders, but include alcoholism, panic attacks and apparently also axis II diagnoses and that which would not agree with Crow's hypothesis. New gene mutations are not required to explain the genetic continuum between the two major psychiatric disturbances: a simple variability of the penetrance function for the transmission system could also the viewed as an exhaustive explanation. It has already been shown that an unlikely number of new mutations would be needed to explain the defective fertility of schizophrenic subjects. We do think that Crow's idea can be interpreted in the light of the biological and hence genetic heterogeneity of Affective and Schizophrenic Disorders. We hope that future studies will reveal more examples that whatever the clinical characteristics, there are subgroups of subjects with similar responses to psycho-active drugs, as has already been shown for the responses to lithium of some subjects with Schizoaffective and Affective Disorders, even though at present, we have methodologies for studying genetically only one spectrum at time.

Furthermore, it is necessary to apply a preliminary exploratory analysis without any assumptions about the origin or nature of the relevant factors affecting the risk for the disease, in order to differentiate those factors that are important variables from those variables that have stochastic meaning and serve only to confuse. Although very important, this kind of analysis has only seldom been applied to Schizophrenic Disorders.

An example is shown in Table 3 which reports our sample arranged according to the variables that have been found to be significantly correlated with the risk for Schizophrenic Spectrum Disorders in relatives. The sex of proband and the sex of the relatives are significant as well as the type of relationship. In relatives of female probands, the frequency of the disease is always higher than in relatives of male probands (8.2% vs. 4.2%), while female relatives show a higher risk than male relatives (7.8% vs. 4.9%).

The sex role in familial epidemiology of Schizophrenic Disorders has not always been correctly evaluated, but our data is indirectly confirmed by the effect of sex on the age at onset distribution. Our previous studies show that the age at onset distribution for females is always later than for males (7).

This fact is more known and it can be attributed to the estrogen effect, which is similar to that of neuroleptics. We also might think that females fall ill, all other factors being equal,

Table 3. Age corrected frequency of schizophrenic spectrum disorders.

	Mild		Severe	
	Male	Female	Male	Female
Fathers				
Observed affected	1	2	6	3
age-corrected total	47.7	42.9	141.7	78.5
observed risk %	2.10	4.66	4.23	3.82
Mothers				
Observed affected	3	1	4	7
age-corrected total	47.8	42.4	141.1	79.7
observed risk %	6.27	2.36	2.84	8.79
Brothers				
Observed affected	2	4	6	10
age-corrected total	63	66.9	109.7	101.6
observed risk %	3.18	5.99	5.47	9.84
Sisters				
Observed affected	1	14	7	6
age-corrected total	63.3	63.5	86.5	78
observed risk %	1.58	22.06	8.09	7.69

Referred from (2).

only if they belong to a family whose risk values are higher, as if this tendency to illness were required to get over a protective factor. On the other hand, even when a Schizophrenic Disorder has been established, females more frequently show less-invalidating forms, such as paranoid compared with mixed or disorganized forms.

The results of adoption and twin studies indicate the role of a specific genetic susceptibility in the interaction with behavioural and biological non-genetic factors, and make it possible to explain familial aggregation of Schizophrenic Disorders. We can try to construct a pattern of inheritance of genetic susceptibility, which takes into account the pointed out effects. This strategy consists of testing fitness of observed risks with theoretical ones, which result from a Single Major Locus or Multifactorial hypothesis with different thresholds of severity according to a sex effect. In simulation studies it has been observed that such patterns are too little sensitive to reject the one or the other inheritance hypothesis. Genetic susceptibility works at a biological level and only inclusion of such variables will make the models selective enough.

When we chose the HLA system as a biological variable, we did not choose it for its strength in the association. We chose it because there have been observed constant associations between allele A1 and more invalidating kind of illness and betwween allele A9 and Paranoid Schizophrenia in ethnically different tested populations. Furthermore, even though we dont yet know the complex relationship between HLA antigens and dopaminergic receptor function, we observed that neuroleptics specifically interfere with the binding of the antibody to HLA A1 on cell membrane and that this interference appears to have clinical consequences. When we studied the sensitivity of the dopaminergic receptors after stimulation with L-DOPA (8) and the changes induced by prolonged treatment with neuroleptics (9), we found an association between the presence of A1 and the inability to change the sensitivity of dopaminergic receptor, as indicated by no increase in the area under the curve of GH.

These observations enabled us to hypothesize that the HLA system is a constituent element of the genetic susceptibility pattern, but it is necessary to determine its relationship with familial risk. This relationship is very clear in our sample, but its demonstration is difficult if we use descriptive figures because the statistical analysis we used, that is the logistic analysis, implies second and third order interactions.

Table 4 shows that HLA alone is not significant, but its interaction with the sex of the relative and the severity of the illness seems to affect significantly the morbidity risk (10).

Starting from another point of view its demonstration becomes more immediate. Absolute values (risk entity) are not so very different for families with A1 positive probands than for families with A1 negative probands (Table 5), and really in the previous analysis HLA typing, as main factor, has not revealed to be positive .

In any case, in 28.5% of the families of A1 positive probands in our sample, there was an affected relative, whereas there was an affected relative in only 17.4% of the families of negative probands. Because the mean frequencies of disease in the relatives of the two groups did not differ significantly, we calculated the probability of finding families with or without affected relatives, including the mean size of the families of the two groups in this calculation. The values obtained were 26.1 percent for the families

Table 4. Logistic analysis of observed risk for Schizophrenic
Spectrum Disorders in relatives of schizophrenic probands.

	Mod.I z-value	Mod.II z-value	Mod.III z-value	Mod.IV z-value	Mod.V z-value
Mean	-19.74*	-19.26*	-18.98*	-16.76*	-16.07*
Type of relationship		-2.48*	-2.34*	-2.50*	-2.53*
Sex of the proband			-2.29*	-2.74*	-2.76*
Sex of the relative			-1.12	-1.13	-1.04
Severity (mild/severe)				-0.86	-0.77
Sex of the proband x severity				-1.91	-2.03*
HLA type					-0.46
HLA type x sex of the relative					-0.99
HLA type x severity					-1.30
Severity x sex of the relative					-0.28
HLA type x severity x sex of the relative					-2.07*
Initial MaxLlik	-620.6	-620.6	-620.6	-620.6	-620.6
Final MaxLlik	-225.3	-222.1	-218.8	-216.6	-213.6

* At p= .05, z-value is 1.96 since z-values are normally
  distributed

Table 5.

MR for Schizophrenic Disorders in families

pb. HLA-A1 pos.   pb. HLA-A1 neg.

6.5%   6.5%

Expected probability of families with secondary cases*

pb. HLA-A1 pos.   pb. HLA-A1 neg.

26.1%   26.8%

Observed % of families with secondary cases*

pb. HLA-A1 pos.   pb. HLA-A1 neg.

28.5%   17.4%

* according to family size.

of the A1 positive probands, which is not very different from the 28.5% observed, and 26.8% for the families of HLA A1 negative probands, which is somewhat higher than the 17.4% observed. We feel that this difference in the latter group was due to a tendency of the secondary cases to group within certain families in such a way as to cause dishomogeneity of the distribution of risk.

The fact that the disease does not run in some HLA A1 positive probands families, is somewhat connected, even if we do not know yet how, with the severity of the illness because significant interaction of HLA system on familial risk included it. Nevertheless, we think we can attempt new genetic analysis on schizophrenic families devided according to HLA system because these groups warrant to be more homogeneous from the biological point of view.

On the other hand, it is very important to verify how to progress in analysing risk factors. If we want to identify a factor which could devide illness frequency among risk subjects, that is to say a factor with discriminant features, first it is necessary to verify its possible statistical significance in the examined population or in a sample selected by fit methods. For example the inference of a different genetic substratum has not any heuristic value in those families whose proband has or has not a pathological VBR; in fact, if the methodology of investigation did not include the validation of the sample of probands selected for TAC investigation, derived informations can neither be generalized nor they can be used as a priori factors in genetic studies which are related to the identification of heterogeneity of the pathology we considered. In such a case, for example, some significant results had to be taken into account (the sex of the proband, z= 3.14; the type of relationship, z= 2.98) to make up a model of familial transmission based on the VBR partitioning criterium, but unfortunately these results are unlikely when applied to the general population.

As far as the particular case of the use of the HLA system as a marker is concerned we have used a different method. Within a great number of families whose probands had all been typed, we have looked for the possible differences of genetic substrata in subjects who showed different HLA structures.

One of our recent studies shows the best fit solution applying the MFP model, with differential thresholds for sex and severity, to the families of our A1 positive and negative probands (11). Parent offspring and siblings correlation values are lower in HLA-A1 negative probands families and indicate lesser degrees of genetic resemblance within the population. However, for both groups of families the MFP model appears to be valid, even if the chi-square values are less good for the negative group than for the positive one, indicating some differences between the groups. In this kind of studies fitness alone does not indicate that what we tested is the mode of inheritance really involved, but it shows that it could be so because risks expected from the model's parameters bear observed risks. MFP model has very wide confidence limits because it has few genetic assumptions and therefore it is very easy for it to be accepted.

In our case, it is important to find negative results, because it enabled us to reject from our analysis the type of inheritance we tested and to generate new hypothesis to test. This is the case of SML model, whose thresholds have also been differentiated according to sex and severity of the illness. In fact among the best fit solutions, the only one which is not refused by the model, that is with a not significant chi-square, is the one for HLA-A1 positive probands families. HLA-A1 negative families refuse it, because neither the best fit solution, that is to say the one with minimum chi-square is included in 5% probability limits.

In other words, observed risks for these families cannot be explained following the hypothesis of a susceptibility biallelic locus. Looking forward to further investigations, some remarks about SML for A1 positive families can be useful. The allele frequency selected by SML model (2.05%) appears to be in accord with theoretical estimates and some experimentally proven values. In fact, different investigators have analyzed hypothetical parametric structures of an SML of susceptibility for schizophrenia, which are reliable for the prevalence of the disorder in their observed populations. The allele frequency, according to Hardy-Weinberg, affects the presence of three different genetypes with a specific distribution of the disease: that is, those with a low rate of phenocopies, a majority of the affected heterozygotics and those with a low rate of affected homozygotics for the sick allele. This also agrees with theoretical studies, that is, only a small fraction of the affected subjects are homozygous and we still dont really understand the meaning of this situation.

The fit of the SML model to the data for our population can be indirectly inferred from the expected estimated value of the prevalence of the disease: the model we applied provides this, estimating the expected population prevalence on the basis of both the parametric set and the raw observed prevalence in families of probands. In this case particularly, the estimated population prevalence agrees quite well with the value we obtained experimentally in a large sample of the Italian general population, especially because these values are not relevant for the schizophrenics as a whole but only for the A1 positive.

We can draw some general conclusions from our results. One

possible interpretation could be that the presence or the absence of specific antigens of the HLA system, which is a biological characteristic of the proband, individuates two different subgroups of families with different degrees of homogeneity with respect to susceptibility for the disease. Selecting our schizophrenic patients according to their HLA types, revealed an interesting genetic structure in families of A1 positive probands, compatible with an SML mode of transmission in which the heterozygote has limited penetrance as elsewhere suggested for schizophrenia as a whole. This is also supported by the non-suitability of the SML hypothesis for the A1 negative probands.

Whether the SML and/or MFP modes of transmission fit Schizophrenic Disorders has long been debated in the past. The validity of the two-allele SML model for Schizophrenic Disorders has been criticized by O'Rourke and Suarez (12), who have carefully reviewed all genetic data for schizophrenia published between until 1982. They ruled out this mode of transmission and felt that a polygenic multifactorial model was important, but it must be remembered that they ruled out the SML hypothesis after they had put together all the data collected by different groups and assumed a common ethiopathogenesis for the Schizophrenic Disorders. Our analysis does not agree with this conclusion suggesting that criteria for grouping based on the recognition of biological heterogeneity may reveal the presence of different genetic patterns to be tested in the future.

REFERENCES

1) Nurnberger J.I. jr., Goldin L.R., De Lisi L.E. and Gershon E.S.: "The Genetics of Schizophrenia" in Schizophrenia: an integrative view. John Libbey - London Paris: 1985, 188-208.

2) Gasperini M., Orsini A., Bussoleni C., Smeraldi E. and Cazzullo C.L.: "Exploratory analisis of the risk in families of schizophrenics" in Schizophrenia: an integrative view. John Libbey - London Paris: 1985, 327-335.

3) Gottesman I.I., Shields J.: "Schizophrenia and Genetics: a twin study vantage point" Academic Press, New York, 1972.

4) Hayward C.: "Affective illness in the families of schizophrenic patients: what does it mean diagnostically?" Arch. Gen. Psychiatry: 1986, 43: 714.

5) K.S. Kendler, A.M. Gruenberg, M.T. Tsuang: Psychiatric illness in first degree relatives of schizophrenic and surgical control patients: a family study using DSM III criteria. Arch. Gen. Psychiatry 42:770-779; 1985

6) G.P. Pope, J.F. Lipinsky: Diagnosis in schizophrenia and manic depressive illness: a reassessment of the specificity of schizophrenic symptoms in the light of current research. Arch. Gen. Psychiatry 35:811-828;1978.

7) L. Bellodi, A. Morabito, F. Macciardi, M. Gasperini, G. Grassi, C. Marzorati-Spairani and E.Smeraldi: Analytic considerations about observed distribution of age of onset in schizophrenia. Neuropsychobiology 8:93-100;1982.

8) E. Smeraldi, F. Brambilla, R Scorza-Smeraldi, C. Maffei, L. Vanzulli and L. Bellodi: Dopamine receptor sensitivity in chronic schizophrenia: a study on HLA system. In: G. Racagni, F. Brambilla and D. De Wied "Progress in Psychoneuroendocrinology" Amsterdam: Elsevier/ North Holland Biomedical Press, 507-519; 1980.

9) E. Smeraldi, L. Bellodi, E. Sacchetti and C.L. Cazzullo: The HLA system and the clinical response to treatment with chloropromazine.
British Journal of Psychiatry 129:486-489,1976a.

10) L. Bellodi, C. Bussoleni, R. Scorza-Smeraldi, G. Grassi, L. Zacchetti and E. Smeraldi: Family study of Schizophrenia: exploratory analysis for relevant factors. Schizophrenia Bulletin 1986; 12:120-128.

11) E. Smeraldi, F. Macciardi, M. Gasperini, A. Orsini, L. Bellodi, G. Fabio and A. Morabito. Genetic modelling in Schizophrenia according to HLA typing.
Clinical Genetics 1986; 30:157-166.

12) D.H. O'Rourke, I.I. Gottesman, B.K. Suarez, J.Rice and T. Reich: Refutation of the general single-locus model for the etiology of schizophrenia.
Am. J. Hum. Genet. 34, 630-649.

13) T.J. Crow: Secular Changes in Affective Disorders and Variation in the Psychosis Gene.
Arch. Gen. Psych.,1986; 43, 1013-1014.

14) E. Smeraldi, M. Gasperini, F. Macciardi, C. Bussoleni,
A. Morabito: Factors affecting the distributions of age
at onset in patients with affective disorders. J. Psychiat. Res.
17:309-317,1983.

4
Life events and schizophreniform disorders
C. Faravelli, S. Pallanti, M.L. Petrulli, S. Stocchetti and R. Goddard

During the last century schizophrenia was considered to be so-
lely a brain disease of entirely endogenous origin, and the aetiology was sought in brain dissections to uncover anatomial abnormalities.
Since 1907 , with the presentation of Jung (1) of the "Psycogenesis
of Mental Disease" and with the hypothesis of Bleuler(2) in 1911
that considered life situations and emotional conflicts as casual
factors in the onset of at least some cases of schizophrenia ,
the role of recent stressful experience became more important.
The discussion which followed fed the important contributions of
the German psychopathologists during the 1950's (3) (4). From this
evolved the empirical work with extensive validation studies and
the adoption of standarized techniques which multiplied and became
more sophisticated during the following decade.
     The question of the relationship between life events and psy-
chosis is of fundamenta importance and the debate on the subject
is by no means concluded even if at the moment the trend is to
admit a relationship between the events and the acute episode
of schizophrenia.The statement presented on p.185 of DSM III(5)
reflects this point of view: "the onset of the affective phase
of schizophrenic disorder, either initially or as an exacerbation
of a preexisting active phase, is often associated with a psychosocial stressor ".
In reality a review of the literature concerning experimental research proposes basically three different models. The first contends
that there is no relationship between life events and schizophrenia,
the second one proposes a stress-released pattern where life events
have a direct causative effect on the onset or reexacerbation of
schizophrenia, while the third one maintains that the schizophrenic style of the life induces a higher risk of life stress.
The first proposition has been supported in the following studies:
Beck & Worthen (6) (1972); Clancy et al (7) (1973); Jacobs et al
(8) (1974); Jacobs & Myers (9) (1976); Lahniers & White (10) (1976)
in these studies no differences between schizophrenics and

controls are verified. It should be noted, however, that normal
controls were only used in the work of Jacobs and Myers (1976),
whereas the others depended on neurological or psychiatric cases.

A similar number of research studies have shown a significant
relationship between life events and schizophrenia for a period of
time varying between a few weeks to several months before the onset
of demonstrable pathology. This proposition has been supported in
the following studies: Stenberg & Durell (11)(1968);Brown & Birley
(12)(1968); Birley & Brówn (13)(1970); Brown et al (14)(1973); Leff
et al (15)(16) (1973, 1983); Leff & Vaughn (17)(1980); Schwartz &
Myers (18)(1977a,1977b); Chung et al (19)(1986).

Our data agree with this latter position. We have evaluated
the life events occurring in the twelve months preceding the acute
onset of the illness in 26 patients meeting the DSM III diagnosis
of schizophreniform disorder. 116 healthy subjects drawn from hospital employees and their relatives were used as controls. Patients
and controls were matched as regard sex, age, social class. The
life events were registered in detail by means of a semi-structured
interview and the records, with patients and controls randomly
mixed, were evaluated by assessors blind as to wether the records
referred to a patient or to a control subject. Life events were
evaluated both using a normative method, based on Paykel's list (20)
and with a contextualmethod, following Brown's procedure (21).

TABLE 1.

	NORMATIVE ASSESSMENT	CONTEXTUAL ASSESSMENT
Schizophrenif. dis. (N=26)	29.9 ± 11.0	11.6 ± 6.4
Healthy controls	15.3 ± 13.8	7.4 ± 6.9
	t = 2.14	t = 2.51
	$p < 0.05$	$p < 0.02$

TABLE 2

NUMBER OF CASES WHO UNDERWENT AT LEAST ONE SEVERE STRESSFUL EVENT

	Yes	No	
Schizophrenif. Dis.	14 (53.8%)	12	26
Healthy controls	31 (26.7%)	85	116
	45	97	142

CHI SQUARE = 7.21, p < 0.01

Patients with Schizophrenic disorder experienced a greater amount of life stress both when this was rated with the normative and the contextual method as can be seen in table 1.

Even considering the number of subjects who underwent at least one severe event (we considered severe the top 20 events of Paykel's list), significantly more patients experienced a severe stressor.(Tab.2) At this point it remains to be established wether such an associa- tion may be regarded as having a causative effect on the aetiology of the illness or not.

The relationship of life events with schizophrenia, in fact, may be viewed in two opposite directions: at one level the disease might be considered as secondary to the increase of life stress, while the alternative position maintais that the events could be due to the disordered behaviour. In fact the subclinical symptoms preceding the manifest onset of schizophrenia and the particular personality described in these subjects may give rise to increased opportunity of experiencing negative life events.

We are faced with problems of both a theoretical and technical nature. Is it theoretically possible to differentiate between eccentricity, schizotypical personality and the prodromal symptoms of an incipient episode of schizophrenic illness? Is it technically possible to devise operational criteria to delineate the precise onset of a relapse?

It is of paramount importance to establish the independence of an event and this is no easy matter. The problem has been rece- ntly addressed in two different ways by separate workers. Paykel (22)(1980) and his colleagues stressed that the independence of the event must be established from the consequences or potential consequences of the psychiatric illness. The other group led by Mac Millan (23)(1986) takes a wider view and considers any effect of the subject's ownactions on the event in the preceding eighteen months should be taken into consideration.

Several studies have attempted to distinguish between dependent

and independent life events and still demonstrate a significant
excess in events preceding schizophrenic illness:
-Brown & Birley (11)(1968):   independent events
                                 excess of events in schizophrenics
-Harris & Peto (24)(1973):   independent events
                                 excess of events in schizophrenics
-Chung et al (19)(1986)   independent events
                                 excess of events in schizoph. dis.
-Birley & Brown (12)(1970   relapses
                                 on drugs 60%   on placebo 31%
-Leff et al (14)(1973)   relapses
                                 on drugs 44%   on placebo 21%

In the first three studies the independent events, i.e. those
events that are out of the subject's control, were evaluated separately. And even in this case schizophrenics experienced a higher
amount of stress. The last 2 papers give indirect support to the
causative model: life events were twice as great in patients who
relapsed while on active treatment compared with patients that relapsed while on placebo. While showing the protective action of drugs,
these findings also indirectly implicate a provoking role of life
events upon recurrences. On the other hand it must be stressed
that in all the studies where only the events that are independent
of the subjects control are considered the excess of events undergone by schizophrenics either disappears or drops dramatically.
Moreover a high correlation between clinical symptoms and life
events has been reported.

When we attempted to single out those events that could be
secondary to the subjects 'behaviours we also noticed a consistent
decrease of the differences, that fell well below the level of
significance. (Table 3)

TABLE 3

_____

NUMBER OF CASES WHO UNDERWENT AT LEAST ONE SEVERE INDIPENDENT
                      STRESSFUL EVENT

	Yes	No	
Schizophrenif. dis.	10 (38%)	16	26
Healthy controls	29 (25%)	87	116
	39	103	142

CHI SQUARE = 1.93, N.S.

36

It seems therefore that both the directions of the relationship
between life events and schizophrenia have to be considered.

Finally, a last point deserves the greatest attention: the
problem of MEANING.

It is obvious that, in order to produce a behavioural response,
a life event must undergo an intrapsychic process, during which
a certain meaning is attributed to the event. It is also evident
that several factors participate in this process of establishing
significance: past experience, previous responses , personality,
context,etc. Ultimately, a meaning that is unique and specific
to that particular individual "here and now" is attached to each
event . This notwithstanding, a portion of meanig that is common
to the others also exists. In other words the personal meaning
may be viewed as the conbination of two distinct portions, one
subjective, individual, irrepeatible, and another that accounts
for the portion of meaning that the subject shares with the others,
such a "mean of the meanings" might be called "collective meaning"
or "group meaning" .

In order to compare these to components we asked a sample of
normal people to evaluate subjectively, on a 10 point rating-scale,
the impact of the events that had actually occurred to them during
the last year.

We have compared this subjective assessment with the normative
evaluation, which may be considered as a measure of the group
meaning. The concordance was decidedly high, since the correlation
coefficient was .81, that means 66% common variance. In other
words 66% of the subjective meaning might be predicted from the
collective meaning.

This is not totally unexpected, since in everyday life, given
a certain situation, we expect other people to react according
to predictable directions.

Yet, in the popular cultural background, what characterises
madness is the unpredictability of behaviour, the abnormability
of the reaction in respect of the event that is at the basis of it.
Even in psychiatry, and in DSM III as well, we often use terms such
as excessive, inappropriate, disproportionate, unmotivated in order
to distinguish normal from abnormal behaviours.
When we repeated the same comparison as before in a group of psychotic
patiens (not necessarily schizophrenics), we found that the agreement between subjective and normative assessments was much less
than in normals. The correlation coefficient was here .51, showing
that only 21% of the subjective meaning was explaind by the collective
meaning. This may be reasonably raise the suspicion that it is the
internal intrapsychic processing of the event that is alterated
in psychosis rather than the actual quality or   severity   of   external   events.

In conclusion, therefore, we feel that the crucial point of the relationship between life events and schizophrenia remains in the mind of patients, while life events have to be regarded as nonspecific cues.

REFERENCES

1.   Jung, C.G. (1907). Uber die Psychologie der Dementia Praecox.
2.   Bleuler, E. (1911). Dementia Praecox or the group of schizophrenias. (New York: International Universities press, 1950).
3.   Schneider, K. (1959). Clinical psychopathology. (New York: Grunej Stratton).
4.   Mayer-Gross, W.; Slater, E. and Roth, M. (1969). Clinical Psychiatry. 3rd ed. (Baltimore: The William & Wilkins Company).
5.   American Psychiatric Association (1980). DSM III Diagnostic and statistical manual of mental disorders. 3rd ed. (Whashington, DC: the Association).
6.   Back, J. and Worthen, K. (1972). Precipitating stress, crisis theory and hospitalization in schizophrenia and depression. Arch. Gen. Psychiat. 26: p.123-129.
7.   Clancy, T.; Crowe, R.; Winokur, G. (1973). The IOWA 500: precipitating factors in schizophrenia and primary affective disorder. Comprehensive Psychiatry, 14, 197-202.
8.   Jacobs, S. ;Prussof, B. and Paykel, E. (1974). Recent life vents in schizophrenia and depression. Psychol. Med., 4, 444-453.
9.   Jacobs, S. and Myers, S. (1976). Recent life events and acute schizophrenic psychosis: a controlled study. J. Nervous and Mental Disease, 162, 75-87.
10. Lahniers, C. and White, K. (1976). Changes of environmental life events and their relationships to psychiatric hospital admissions. J; Nervous and Mental Disease, 163, 154-157.
11. Steinberg, H. and Durell, J. (1968). A stressful social situation as a precipitant of schizophrenic symptoms: an epidemiological study. Brit. J. Psychiat., 114, 1097-1105.
12. Brown, G.W. and Birley, J.L. (1968). Crises and life changes and the onset of schizophrenia. J. Health and Social Behaviour, 9 203-214.
13. Birley, J.L.T. and Brown, G.W. (1970). Crisis and life changes preceding the onset or relapse of acute schizophrenia: clinical aspects. Brit. J. Psychiat., 116, 327-333.
14. Brown, G.W.; Harris, T. and Peto, J. (1973). Life events and psychiatric disorders: part. II. Nature of the causal link. Psychological Medicine, 3, 159-176.

15. Leff, J.P.; Hirsch, S.R.; Gaind, R.; Roth, P. and Stevens, B.S. (1973). Life events and maintainance therapy in schizophrenic relapse. Brit. J. Psychiat. 123, 659-660.

16. Leff, J.; Knipers, L.; Berkowitz, R.; Vaughn, C. and Sturgeon, D. (1983). Life events, relatives' expressed emotion and maintainance neuroleptics in schizophrenia relapse. Psychological Medicine, 13, 799-806.

17. Leff, J.and Vaughn, C. (1980). The interaction of life events and relatives expressed emotion in schizophrenia and depression neurosis. Brit. J. Psychiat., 136, 146-153.

18. Schwartz, C.C. and Myers, J.K. (1977). Life events and schizophrenia: II impact of life events on symptom configuration. Arch. Gen. Psychiatry, 34, 1242-45.

19. Chung, R.K.; Langeluddecke, P. and Tennant, C. (1986). Threatening life events in the onset of schizophrenia, schizophreniform psychosis and hypomania. Brit. J. Psychiatry, 148, 680-685.

20. Paykel, E.S.; Prussof, B.A. and Uh Lenhtuh, E.H. (1971). Scaling of life events. Arch. Gen. Psychiat., 25, 340-347.

21. Brown, G.W. (1974). Meaning measurement and stress of life events. In: Dohrenwend, B.S. and Dohrenwend, G.P.: stressful life events: their nature and effects. Wiley, New York.

22. Paykel, E.S. (1980). Recall and reporting of life events. Arch. Gen. Psychiatry, 37, 485.

23. Mac Millan, J.F.; Gold, A.; Crow, A.L. et al (1986). Expressed emotion and relapse. Brit. J. Psychiat., 148, 133-143.

24. Brown, G.W.; Harris, T. and Peto, J. (1973). Life events and psychiatruc disorders: part. II. Nature of the causal link. Psychological Medicine, 3, 159-176.

5
Hypotheses on the seasonality of schizophrenic births

E.F. Torrey

"To everything there is a season and a time to every purpose under heaven".

Ecclesiastes, Chapter 3

INTRODUCTION

The seasonality of births of persons who later develop schizophrenia is one of the most firmly established facts about this disease. At least 44 studies in 18 different countries including Italy have confirmed that an excess number of schizophrenics are born in the winter and spring months [1-4]. According to one recent review "in studies with a sample size of over 1,500 the excess ranges from 0 to 20 percent with a mean of about 10 percent" [3]. The trend is more clearly evident in studies of northern hemisphere countries, but the trend in southern hemisphere countries is in the same direction corresponding with their reversed seasons. It is also now widely accepted that the seasonal excess of schizophrenic births is real and not a statistical artifact based on age-incidence or age-prevalence effects [3].

Two other aspects of the seasonality of schizophrenic births appear to be firmly established. One is that there may be regional differences in the seasonality within a country, e.g., it is more pronounced in northern Sweden than in southern Sweden [5], and more pronounced in northern states in the United States than in southern states [6]. The other is that the seasonality of schizophrenic births is not necessarily static but may shift over time as has been shown in studies in Japan [7], England [8], and the United States [9].

Other aspects of the schizophrenic seasonality question are less conclusively settled. The bulk of evidence suggests that seasonality of births is not found among patients with bipolar disorder [2,3] but others have disputed this conclusion [10]. Neuroses and personality disorders also apparently do not show a seasonal birth pattern but this is also not conclusive [11,12]. There have also been attempts to divide schizophrenic patients into groups to determine whether such groups have more or less seasonality of births. The data so far is inconclusive on groups divided by sex, race, family history for schizophrenia, clinical subtype, positive or negative symptoms, marital status, age of onset, and length of hospitalization [1-3,11,13,14].

41

It should also be noted that schizophrenia is not unique in have a seasonality of births. Many congential anomalies and chromosomal dysjunctions show a seasonality of births of those affected. Cleft-lip cleft-palate, hypospadias, positional foot defects, Down's syndrome, Turner's syndrome, and Klinefelter's syndrome all occur more commonly in spring births; patent ductus arteriosus occurs in the summer; the Prader-Willi syndrome has a peak in fall births; and congenital dislocation of the hip and anencephaly are found most frequently among winter births [15-19] . It has also been noted in some studies that uterine bleeding, spontaneous abortions and stillbirths occur more commonly in winter and spring births [11] . Among other diseases there have been suggestions that patients with goitre, rheumatoid arthritis, and perhaps insulin-dependant diabetes mellitus may have a seasonality of their births [5] ; in the latter a study in Iceland attempted to link the seasonality to the seasonal ingestion of nitrosamines in cured meats [20] .

HYPOTHESES REGARDING CAUSE OF SEASONALITY

The most interesting question regarding the seasonality of schizophrenic births is what causes it. Eight hypotheses have been put forward as possible explanations; they will be examined in order of ascending credibility.

(1) Exaggerated seasonality of general births

It has been known for many years that general births show a slight seasonality, and it has been suggested that the schizophrenic birth seasonality is simply an exaggeration of those general birth trends. It is not known what is responsible for the seasonality of general births but factors such as seasonal female endocrinological cycles, climate, increased spontaneous abortions and stillbirths, and seasonal variations in frequency of intercourse due to cultural factors (e.g., holidays, peak seasons for weddings, harvest, migrant labor) have all been postulated. In Europe the peak for general births occurs in the spring with a smaller secondary peak in September (nine months after the Christmas holidays), and the spring peak coincides closely with the excess schizophrenic births peak. In the United States, however, there is no spring peak for general births but rather a marked fall peak [21] . The peak for schizophrenic births, on the other hand, is in the spring just as it is in European countries. This would strongly suggest that the seasonality of schizophrenic births is not related to the seasonality of general births.

(2) Genetic fitness

The genetic fitness hypothesis states that the theoretical schizophrenic genotype confers increased robustness onto the newborn giving it, for instance, greater resistance to infections [22] . For pre-schizophrenic babies born during the winter or spring months when infections are more common, the genetic constitution of the child would increase its survival rate over non-pre-schizophrenic babies and thereby produce a winter-spring seasonality of schizophrenic births. This hypothesis cannot be ruled out but seems unlikely. Given the greater exposure of newborns to infections a century ago, if this mechanism was responsible for the seasonality of schizophrenic births then one could have expected a greater seasonality in the past than is found now; studies in Japan fail to show any difference in amplitude between the nineteenth and twentieth centuries [7] . This hypothesis would further lead to the expectation that schizophrenics with a family history of the disease (thus carrying the schizophrenic genotype) should have a greater seasonality of births compared with schizophrenics with a negative family history; the evidence to date, although not conclusive, is in exactly the opposite direction [3] .

(3) Procreational habits

The procreational habits hypothesis posits that men and women who them-selves have a schizophrenic genotype conceive more often during the summer months and thus produce an excess of late winter and spring births of children whose chances of developing schizophrenia are considerably higher than that of the general population. One variant of this theory is that "in the summer people wear fewer clothes in bed, and that a schizoid spouse is more likely then to notice his(or her) co-spouse there and accordingly to initiate sexual behavior" [23] . Another variant was the suggestion that, until recently, the only time male and female patients in mental hospitals could mingle socially was on the hospital grounds during the summer and that this mingling led to an excess schizophrenic births nine months later [24] .

The procreational habits hypothesis would logically predict that schizo-phrenics with a family history of the disease would show a greater seasonality of births than those without a family history; as mentioned above the evidence to date is in the opposite direction. It would also predict that climates with a greater difference in summer and winter temperatures (e.g., Sweden) should show greater seasonality of schizophrenic births than climates with less differences (e.g., Philippines); although the overall evidence leans somewhat in that direction it is not yet clearly established.

Finally this hypothesis would predict that the siblings of schizophrenics should show a seasonality of births similar to their schizophrenic siblings since their parents' procreational habits would affect all children equally. Two studies of 302 and 272 siblings did show results in this direction but neither achieved statistical significance [25,26] . On the other hand three other studies of 777, 1,322, and 2,501 siblings failed to find a seasonality of siblings births similar to that of schizophrenics; one even found a slight deficit of January to March births among the siblings [27-29] . The analysis is complicated by the fact that families plan pregnancies to space children, and that the winter or spring birth of a child destined to become schizo-phrenic will influence the timing of the birth of the next child because the parents may want the children to be a specified period apart (e.g., about two years). To validly test this hypothesis, therefore, the birth pattern of older siblings only should be examined; the author currently has such research underway.

(4) Temperature effect

This hypothesis states that the temperature either at the time of conception, during pregnancy, or at the time of birth has an effect on the developing fetus in a way which increases the liklihood of it developing schizophrenia. Pasamanick and Knobloch, for example, claimed that hotter summers cause more protein deficiency and this in turn produces more schizophrenia in children conceived during the summer months [30] . One group of researchers were able to confirm a relationship between the temperature at conception and the birth of schizophrenics [31] but four other studies which examined temperature and schizophrenic birth patterns did not find any significant relationships [2] . The hypothesis would further suggest that the seasonality of schizophrenic births should be related to geographical factors (e.g., more seasonality in hotter climates) but there is no evidence to support this theory. Still another problem with the temperature effect hypothesis is that the peak seasonality of schizophrenic births does not necessarily occur during the coldest months. In Ireland, for example, the peak seasonality is during the second quarter of the year (April, May, June) rather than during the first (January, February, March) [32] . A variant of the temperature effect is one involving amount of sunshine; this was put forth, for example, as an explanation for anencephaly in Scotland [33] but has never been explored for schizophrenia.

(5) Perinatal complications

As mentioned above, it is known that many complications of pregnancy and birth, including spontaneous abortions and stillbirths, have a seasonal occurrence similar to that found in schizophrenic births. The perinatal complications hypothesis states that seasonal adverse perinatal events lead to brain damage in the affected fetus which in turn may cause schizophrenia.

There is some evidence in support of this hypothesis. Videbech et al. found a correlation between schizophrenic births and stillbirths in Denmark [34] . And Kinney and Jacobsen, analyzing data from a very small number (N=34) of subjects, reported that a history of brain damage was found more often in schizophrenics born in the winter months [35] . This hypothesis would also predict that schizophrenics born in the winter and spring months would show more evidence of neuropathological damage on such measures as computerized tomography (CT) scans; preliminary evidence from an Italian study showed precisely this [4] .

At the same time there is an element to this hypothesis which is unsatisfying. It fails to identify the cause of the seasonal adverse perinatal events, merely posits their occurrence and in that sense the hypothesis is more a description of a possible intermediate step in the etiological sequence leading to the seasonality of schizophrenic births than it is a complete explanation. Seasonal nutritional deficiencies, environmental contaminants or infectious agents could all theoretically be the primary etiological agent leading to the perinatal complications and ultimate seasonality of birth.

(6) Nutritional Deficiences

Since the nutrition of pregnant mothers and infants varies in different seasons, it is reasonable to hypothesize that such factors, by causing brain damage, could account for the seasonality of schizophrenic births. Pasamanick and Knobloch [30] proposed that protein deficiency occurs most commonly in the summer months, the first trimester for winter births, and that this produces both more schizophrenic and more mentally retarded individuals.

Other nutritional theories proposed have included a deficiency of vitamin C or vitamin K [5] during the winter months; the latter is associated with an increased bleeding tendency (hemorrhagic disease of the newborn), which theoretically could cause brain damage. Another observer suggests vitamin D deficiency (because of less sunlight in northern winter climates) and/or hypocalcemia at the time of birth as possible causes.

The nutritional deficiences hypothesis is logical and certainly possible, but almost no studies have been done which either support or refute it. Since

44

there are differences in the degree of schizophrenic birth seasonality in different geographic areas (e.g., greater in northern states than southern states in the U.S.) such hypotheses could be tested. Given the improved nutrition available now compared with earlier years (e.g., the nineteenth century) one would also predict that the seasonality of schizophrenic births would have been more pronounced in the past than it is now. Dalen [5] suggests that this has occurred in Sweden, but in Japan [7] this is not the case.

(7)   Environmental contaminants

A wide range of agents might theoretically act as seasonal evnironmental contaminants, affecting the fetus or newborn and possibly leading to schizophrenia. One example is lead intoxication which is known to occur seasonally [36]. In a study of plasma magnesium in schizophrenia it was also found that levels of magnesium in the blood varied seasonally [37]. Organic solvents are known to cause defects in the central nervous sytem in fetuses whose mothers are exposed during pregnancy [38] and such exposure could occur on a seasonal basis for occupational reasons. Toluene, for example, is thought to be capable of causing "an irreversible schizophreniform psychosis" following chronic exposure [39] while others have described schizophrenic-like psychosis following exposure to phosphate-ester insecticides which are used seasonally [40].

Pharmacological agents might also be taken by pregnant women on a seasonal basis (e.g., aspirin during flu season) and affect the developing CNS in unknown ways to produce later schizophrenia. Researchers have also hypothesized about the possibilities that the seasonal exposure of pregnant women to the nitrosamines in smoked meat in Iceland might cause insulin-dependent diabetes [20], or that the seasonal exposure to the alkaloids in blighted potatoes might produce anencephaly and spina bifida [41]. Another type of environmental contaminant is radioactivity; a study in Alberta attempted to correlate the seasonal occurrence of congenital anomalies there with radioactivity carried to earth by rainfall [42].

There is no evidence than any of these environmental contaminants, or any others, play a role in the cause of schizophrenic births seasonality. Studies in this area, however, are virtually nonexistant. Given the sensitivity of the developing CNS to minor changes in its chemical environment, and the fact that many of these contaminants occur seasonally, they warrant a much closer look.

(8)   Infectious agents

Viruses are attractive candidates to explain the seasonality of schizophrenic births because many of them are known to have a seasonal pattern, to be neurotropic, and to remain latent for 20 years or longer. Given the continuing discovery of other infectious agents, however (e.g., slow "viruses", prions) these should also be considered. At least 15 viruses are known which cause infections in fetuses; rubella, measles and varicella-zoster occur most commonly in the winter and spring months. Rubella, for example, has a marked spring seasonal occurrence which produces a sharp seasonal peak of newborns with congenital rubella born in October through January. Other viruses such as the Ebstein-Barr virus occur much more evenly throughout the year with a slight peak in the spring and the fall.

Three attempts have been made to look for correlations between schizophrenic birth seasonality and infectious agents. In Finland it was found that a disproportionate number of schizophrenics had been in utero during the 1957 influenza epidemic there [43] . In Minnesota the variation in five bacterial (diphtheria, pneumonia, scarlet fever, pertussis, typhoid fever) and three viral (influenza, measles, polio) diseases were compared with the seasonal pattern of schizophrenic births between 1915 and 1959; schizophrenic birth seasonality was found to be significantly greater "in the years directly following those marked by high levels of infectious disorders" (p< .05) most marked for diphtheria, pneumonia and influenza [44] . Another study examined the relationship between reportable viral illness and schizophrenic births in Connecticut and Massachusetts and reported statistically significant correlations for measles, polio, and varicella-zoster [45] .

A viral explanation for schizophrenic birth seasonality is also attractive because of its explanatory power for many other aspects of the disease. It has long been known that viral encephalitis, especially when caused by viruses in the herpes family, may mimic schizophrenia in its early stages. And many of the neuropathological lesions found in schizophrenia, including CT scan findings, are consistent with a viral etiology. Viruses could also explain epidemiological aspects of this disease, including its marked prevalence differences in different areas of the world. The relapsing-remitting course of schizophrenia is similar to multiple sclerosis, another CNS disease with a suspected viral etiology. Dermatoglyphic and perinatal abnormalities found in some schizophrenic patients could be accounted for by an in utero viral infection, and some viral infections alter dopamine levels, which are widely believed to be affected in schizophrenia. Finally, some viruses can become incorporated into genes and transmitted genetically, while others only affect individuals with a genetic predisposition to them; thus genetic aspects of schizophrenia can also be accounted for by a viral hypothesis [1] .

SUMMARY

The seasonality of schizophrenic births is one of the firmly established facts about this disease. Different hypotheses to account for this phenomenon are examined and it is concluded that the most likely explanations are infectious agents, environmental contaminants, or nutritional deficiencies.

REFERENCES

1.   Torrey, EF and Kaufmann CA (1986). Schizophrenia and neuroviruses. In: Nasrallah, HA and Weinberger, DR (eds.) "The Neurology of Schizophrenia". (Amsterdam: Elsevier)
2.   Bradbury, TN and Miller, GA (1985). Season of birth in schizophrenia: A review of evidence, methodology and etiology. Psychol Bull, 98, 569
3.   Boyd, JH, Pulver, AE and Stewart, W (1986).   Season of birth: Schizophrenia and bipolar disorder. Schiz Bull, 12, 173
4.   Cazzullo, CL, Caputo, D, Bellodi, L, Maffei, C, Ferrante, P, Bergamini, F, Landini, MP and La Placa, M (1984). Schizophrenia: An epidemiological, immunological and virological approach.   Presented at First World Conference on Virus Disease and Mental Health, Montreal, 1984
5.   Dalen, P (1975). "Season of Birth: A Study of Schizophrenia and Other Mental Disorders". (Amsterdam: North-Holland)

6.   Torrey, EF, Torrey, BB and Peterson, MR (1977).   Seasonality of schizophrenic births in the U.S. Arch Gen Psychiat, 34, 1065
7.   Shimura, M and Miura, T (1980).   Season of birth in mental disorder in Tokyo, Japan, by year of birth, year of admission and age at admission.   Acta Psychiat Scand, 61, 21
8.   Hare, EH (1978).   Variations in the seasonal distribution of births of psychotic patients in England and Wales. Brit J Psychiat, 132, 155.
9.   Torrey, EF and Torrey, BB (1979).   A shifting seasonality of schizophrenic births. Brit J Psychiat, 134, 183
10.   Ulwelling, W (1985).   Winter births and seasonal affective disorder. Arch Gen Psychiat, 42, 105
11.   Torrey, EF (1980).   "Schizophrenia and Civilization". (New York: Jason Aronson)
12.   Watson, CG, Tilleskjor, C, Kucala, T and Jacobs, L (1984).   The birth seasonality effect in nonschizophrenic psychiatric patients.   J Clin Psychol, 40, 884
13.   Opler, LA, Kay, SR, Rosado, V and Lindenmayer, JP (1984).   Positive and negative syndromes in chronic schizophrenic inpatients.   J Nerv Ment Dis, 172, 317
14.   Lo, CW (1986).   Season of birth of schizophrenics in Hong Kong.   Brit J Psychiat, 147, 212
15.   Jongbloet, PH (1971).   Month of birth and gametopathy.   Clin Genet, 2, 315
16.   Bailer, JC and Gurian, J (1965).   Congenital malformation and season of birth: A brief review. Eug Quart, 12, 146
17.   Butler, MG, Ledbetter, DH and Mascarello, JT (1985).   Birth seasonality in Prader-Willi syndrome. Lancet, 2, 828
18.   McKeown, T and Record, RG (1951).   Seasonal incidence of congenital malformations of the central nervous system. Lancet, 1, 260
19.   Wehrung, DA and Hay, S (1970).   A study of seasonal incidence of congenital malformations in the United States. Brit J Prev Soc Med, 24, 24
20.   Helgason, T and Jonasson, MR (1981).   Evidence for a food additive as a cause of ketosis-prone diabetes. Lancet, 2, 716
21.   Cowgill, UM (1966).   Season of birth in man: Contemporary situation with special reference to Europe and the southern hemisphere. Ecology, 47, 614
22.   Hare, EH and Price, JS (1968).   Mental disorder and season of birth: Comparison of psychoses and neurosis. Brit J Psychiat, 115, 533
23.   James, WH (1978). Seasonality in schizophrenia. Lancet, 1, 664
24.   Dawson, DF (1978).   An explanation for seasonality of births in schizophrenia. Am J Psychiat, 135, 1434
25.   Hare, EH (1976).   The season of birth of siblings of psychiatric patients. Brit J Psychiat, 129, 49
26.   McNeil, T,. Kaij, L and Dzierzykray-Rogalska, M (1976).   Season of birth among siblings of schizophrenics. Acta Psychiat Scand, 54, 267
27.   Buck, C and Simpson, H (1978).   Season of birth among the sibs of schizophrenics. Brit J Psychiat, 132, 358
28.   Larson, CA and Nyman, GE (1976).   Birth month of schizophrenics and their sibs. IRCS Med Sci, 4, 56
29.   Torrey, EF. Birth seasonality of schizophrenic sibs. Unpublished data
30.   Pasamanick, B and Knobloch, H (1961).   Epidemiologic studies on the complications of pregnancy and the birth process.   In: Caplan, G (ed.) "Prevention of Mental Disorders in Children". (New York: Basic Books)

31.   Templer, DI and Austin, RK (1980).   Confirmation of relationship between temperature and the conception and birth of schizophrenics.   J Orthomol Psychiat, 9, 220

32.   O'Hare, A, Walsh, D and Torrey, F (1980).   Seasonality of schizophrenic births in Ireland.   Brit J Psychiat, 137, 74

33.   Record, RG (1961).   Anencephalus in Scotland.   Brit J Prev Soc Med, 15, 93

34.   Videbech, T, Wecke, A and Dupont, A (1974).   Endogenous psychoses and season of birth.   Acta Psychiat Scand, 50, 202

35.   Kinney, DK and Jacobsen, B (1978).   Environmental factors in schizophrenia: New adoption study evidence and its implications for genetic and environmental research.   In: Wynne, LC, Cromwell, RL and Matthysse, S (eds.) "The Nature of Schizophrenia".   (New York: John Wiley)

36.   Hunter, JM (1978).   The summer disease: Some field evidence on seasonality in childhood lead poisoning.   Soc Sci Med, 12, 85

37.   Yassa, R, Nair, NPV and Schwartz, G (1979).   Plasma magnesium in chronic schizophrenia.   Internat Pharmacopsychiat, 14, 57

38.   Holmberg, PC (1979).   Central nervous system defects in children born to mothers exposed to organic solvents during pregnancy.   Lancet, 2, 177

39.   Goldbloom, D and Chouinard, G (1985).   Schizophreniform psychosis associated with chronic industrial toluene exposure: Case report.   J Clin Psychiat, 46, 350

40.   Gershon, S and Shaw, FH (1961).   Psychiatric sequelae of chronic exposure to organo-phosphorus insecticides.   Lancet, 1, 1371

41.   Renwick, JH (1972).   Anencephaly and spina bifida are usually preventable by avoidance of a specific but unidentified substance present in certain potato tubers.   Brit J Prev Soc Med, 26, 67

42.   LeVann, LJ (1963).   Congenital abnormalities in children born in Alberta during 1961.   Canad Med Ass J, 89, 120

43.   Mednick, SA, Machon, RA, Huttenen, M and Bonett, D.   The 1957 Helsinki type A-2 influenza epidemic and adult schizophrenia.   Submitted for publication

44.   Watson, CG, Kucala, T, Tilleskjor, C and Jacobs, L (1984).   Schizophrenic birth seasonality in relation to the incidence of infectious diseases and temperature extremes.   Arch Gen Psychiat, 41, 85

45.   Torrey, EF, Rawlings R and Waldman, IN.   Schizophrenic births and viral diseases in two states.   Submitted for publication.

6
The retrovirus/transposon hypothesis of schizophrenia

T.J. Crow

INTRODUCTION

A role for genes in the aetiology of schizophrenia is established by twin and adoption studies but the size and nature of the contribution remain in doubt. Only 10% of patients have an affected parent, and concordance in monozygotic twins is probably no greater than 50%. For these reasons most workers have concluded that genes are only one of several contributors to aetiology. Either genes predispose to environmental pathogens or there are both genetic and non-genetic forms of illness.

To identify the non-genetic aetiological factors is the difficulty. Most of the common classes of aetiological agent can be ruled out. Thus trauma, toxins and neoplasia do not seem to play a role. The known pathology and epidemiology of the disease are inconsistent with a contribution from such factors as ordinarily understood. This leaves immunity and infection as prime suspects, and immune dysfunction is often secondary to infection.

1. Origins of the viral hypothesis

The concept of schizophrenia as an infectious disease goes back to the middle of the nineteenth century, particularly in relation to cases of folie à deux. That the disease might be caused by a virus was first seriously considered by K. Menninger and E. Goodall in the aftermath of the 1918 epidemic of influenza and the outbreak of encephalitis lethargica which succeeded it and may have been related. Post-influenzal and post-encephalitic psychoses were seen which had schizophrenia-like features. Subsequently there have been reports of schizophrenic illnesses associated with viral infections, although relationships to identified viruses have not been established (Torrey and Peterson, 1976 [1]; Crow, 1978 [2]). It is clear that some of the characteristic symptoms of schizophrenia may be provoked by infection although this appears to be a relatively unusual, or at least infrequently recognised event. The question arises whether schizophrenia itself could be caused either by an atypical response to a common pathogen or by a hitherto unrecognised virus. Genetic factors predispose to some infections e.g. polio and tuberculosis. Schizophrenia could be due to a virus to which certain individuals are predisposed by their genes.

Recent discussions of epidemiology are relevant. It has been argued that significant changes over time (e.g. an increase in the course of the nineteenth century, Hare, 1983 [3]) have taken place, and also that there are variations across the populations of the world (Torrey, 1980 [4]). Such variations are consistent with spread in the manner of an epidemic.

2. Refutation of the contagion theory

If the disease is transmitted from one individual to another this should be seen within families as well as at the population level. Some findings in family studies can be interpreted as consistent with contagion (Crow, 1983 [5]). For example concordance rates are generally reported as higher in same-sex than opposite-sex pairs of relatives, and in dizygotic twins than in siblings. In these cases it could be that higher concordance rates are seen in pairs of relatives who are likely to be in closer physical contact. Similarly in Abe's analysis of onset of illness (Abe, 1969 [6]) in monozygotic twins the second twin was at increased risk in the 2 years following illness onset in the first twin, the increase in risk being confined to pairs who were together at this time.

A critical test of the contagion hypothesis can be conducted in pairs of siblings. An analysis of age of onset in five reported series reveals a consistent tendency for age of onset to be earlier in younger siblings (Crow and Done, 1986 [7]).

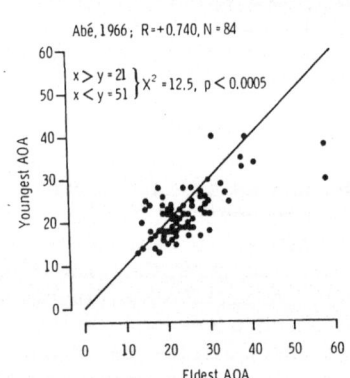

FIGURE 1   Age on (first) admission (AOA) in elder sibling plotted against AOA in younger sibling (from Crow and Done, 1986 [7]).

Three explanations must be considered:-
(i)   that the disease is transmitted from one sibling to the other
or that both are exposed to an exogenous pathogen at the same point
in time (the "contagion" or "horizontal transmission" hypothesis).
(ii)   that the disease is detected earlier when it is already known
to be present in the family ("early detection")
(iii) that because the data are collected at a particular point in
time (e.g. close to onset of illness in one sibling) a bias enters
excluding later onsets of illness in younger siblings ("ascertainment bias").
        The first two hypotheses predict that the shift to younger age
of onset will be seen in pairs in which the elder sibling is ill
first. However this appears not to be the case (figure 2).

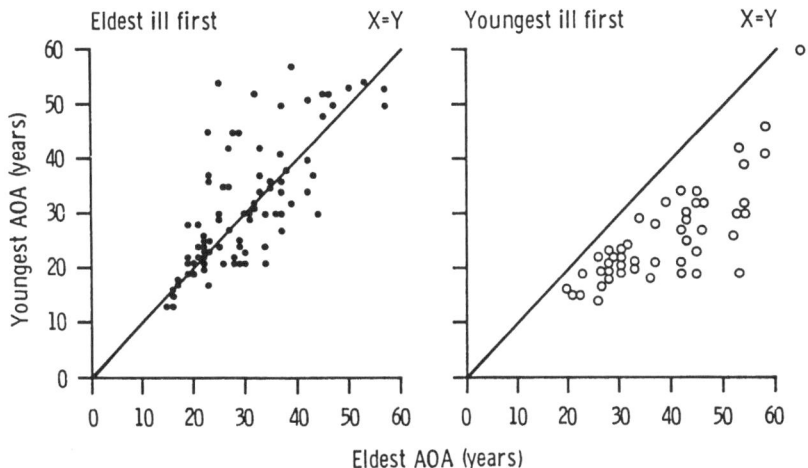

FIGURE 2   Age on admission (AOA) in elder sibling plotted against
            AOA in younger sibling according to which sibling is ill first
            (from Crow and Done, 1986 7).

        In elder sibling ill first pairs mean age of onset in the
siblings is approximately equal. Ascertainment bias rather than
contagion or early detection accounts for the trend toward younger
age of onset in younger sibling.
        When this bias is taken into account (as in the elder-sibling
ill first sub-sample) age of onset remains highly correlated
between siblings but apparently uninfluenced by actual time of
onset. The findings not only rule out contagion (at least in adult
life) but also discourage the concept that onset of illness is
influenced at all by environmental factors of the type that occur
at a defined point in time. It is as though a clock is ticking
away within the predisposed individual to ensure not only that he
will suffer from the illness but that it will occur at a predeter-
mined age.

The findings do not rule out an environmental influence in the prenatal period. Such an influence is suggested by the season of birth effect (Dalen, 1975 [8]) although whether this reflects an event at birth or earlier is uncertain. It is also unclear how events at this stage of life can influence the onset of illness 20 or 30 years later.

An hypothesis (Crow, 1984 [9], 1986 [10]) that attempts to account for the season of birth effect as well as onset of illness in adult life is as follows:-

That the disease results from the expression of a pathogen (e.g. a retrovirus or other mobile genetic element) in the human genome which either is inherited in its pathogenic form direct from an affected or predisposed parent or acquires this form as a result of a genetic rearrangement (e.g. a transposition or replication event) occurring early in development (e.g. at meiosis or in early embryogenesis).

According to this hypothesis the season dependence of such events is responsible for the deviation of birth dates of psychotic patients from expectation for the general population.

3. Location of the element in the genome

Evidence from electrophysiological, CT scan and post-mortem studies suggests that the schizophrenic process has a predilection for the left or dominant hemisphere (Crow, 1986 [10]). This is difficult to understand if the disease is due to an exogenous pathogen but might be explained if the pathogen is integrated in association with a gene which is expressed preferentially in the left hemisphere. An asymmetry of cerebral structure is present in the planum temporale, this region being of greater extent on the left side of the brain in most individuals. The asymmetry develops early. Presumably it is due to the differential expression of a growth factor (the "cerebral dominance gene"). Selectivity for the left hemisphere might be explained if the "virogene" were integrated close to the cerebral dominance gene and expressed in association with it.

4. Nature of the element

Agents of the retrovirus class sometimes become associated with cellular growth factors and may be responsible for aberrations of cellular development. Other types of retrotransposon (e.g. LINE elements) may do likewise.

A number of types of retrovirus sequence have been identified in the human genome. These include the HLM-2 group of sequences identified by hybridisation to mouse mammary tumour virus (Callahan et al. 1985 [11]) and another group identified by hybridisation to retroviral sequences in an African green monkey genomic library (Steele et al. 1984 [12]). There are approximately 50 copies of each of these sequences in the human genome. About 800 to 1,000 copies of a retrovirus-like sequence originally identified in the β-globin gene cluster (Mager and Henthorn, 1984 [13]) are also present. Whether any of these sequences are transcribed in normal or pathological circumstances remains to be established. In addition a 2-3Kb transposon-like element has been found which is also present in extrachromosomal circular DNA (Paulson et al. 1985 [14]).

5. Mode of transmission

If the determinants of schizophrenia are, as argued above, primarily genetic the persistence of the disease at a relatively high prevalence in the populations of the world in the face of reduced fertility in affected individuals remains to be explained.

One possibility is that the gene confers advantages on individuals who carry the gene but are unaffected by the disease. However such individuals and the nature of the advantage have not been identified. Another possibility arises if the pathogen is an element which is mobile within the genome. As pointed out by Hickey (1982) [15] such an element might replicate between haploid genomes to circumvent the laws of Mendelian genetics, and spread within the population even though it reduced fitness. The agent would thus be horizontally transmitted not at a time close to disease onset but at gamete formation. It is not clear however whether an equilibrium state such as is observed in schizophrenia could be set up in this manner.

An alternative explanation is that schizophrenia is but a part of a wider genetic spectrum. There is a case that schizophrenia and manic-depressive psychosis rather than representing distinct disease entities are points on a continuum extending from unipolar through bipolar affective illness to schizo-affective psychosis and schizophrenia with increasing severities of defect. The case is argued on the basis of the failure of attempts to demonstrate a bimodal distribution of scores on a discriminant function of symptoms, on the overlap in family studies of schizo-affective psychosis with both of the prototypical psychoses, and on evidence for an excess of individuals with schizo-affective disorders and schizophrenia amongst the descendants of patients with affective disorders (Crow, 1986 [16]). The hypothesis generates the prediction that the gene responsible for psychosis is itself variable, the presence of the gene in a pathogenic form increasing risk of disease of greater severity. Such increases in severity might be due to replications within the genome, e.g. the generation of tandem repeats.

The concept leads to the suggestion that the reason for the persistence of the schizophrenia gene lies elsewhere on the continuum. There is evidence for example for an association between affective illness and creativity (Jamison, 1986 [17]) and also for particular achievement in the relatives of patients with schizophrenia (Karlsson, 1984 [18]). In a study in Norway (Noreik and Odegaard, 1966 [19]) amongst those with occupations in the professional services admission for manic-depressive psychosis was increased and for schizophrenia decreased with respect to the general population. This suggests the relationship between achievement and affective illness may be closer than with schizophrenia. However while fertility is certainly decreased in schizophrenia it is far from clear that it is increased in affective illness. If affective illness and schizophrenia are genetically related the problem of persistence is moved back but not yet solved.

The questions of persistence and possible genetic advantage may be related to the problem of age of onset. As already noted age of onset of schizophrenia is earlier than that of affective disorder, and of bipolar illness than of unipolar illness. Thus there is an inverse relationship between age of onset and severity of illness along the continuum. However it is also recognised that typical psychotic illnesses seldom have an onset earlier than puberty. It is as though an evolutionary feedback mechanism were hunting for a time of onset in the period following puberty. Perhaps this is a time of particular cerebral growth.

Asymmetrical brain growth is a recent evolutionary development presumably related to the specifically human capacity for speech and perhaps to other intellectual abilities. Its integration with other aspects of brain growth must be critical for optimal cerebral development. One might suppose that the genetic mechanisms respons- ible require an intrinsic timing device and also are relatively variable between individuals, such variability applying both to the magnitude of the effect and its timing. Pathological effects could arise from inappropriately great enhancements of growth and their severity might be determined by the stage of development at which they occur.

The findings of a recent CT scan study (Colter et al. in preparation [20]) illustrate a possible sequence. Asymmetries can be observed on CT scan in the temporo-occipital region in the normal brain, the width of the brain being greater on the left. Overall comparison of schizophrenics with other psychiatric patients reveals no difference in the asymmetries. However patients with early onset (less than 25 years) have diminished, and with late onset (over 25 years) increased asymmetries, the difference between the groups being significant at the 1% level. The findings rein- force the concept of the disease as a disorder of cerebral asymmetry development. One interpretation is that early disease expression (perhaps as viral particles) causes arrest of develop- ment while later onset is associated with enhanced growth, which also eventually ends in pathology.

6. <u>Summary</u>

The concept that schizophrenia is a disease induced in genetically- predisposed individuals by an exogenous virus has been tested in pairs of affected siblings. Onset is determined by age and not by common exposure to an environmental pathogen. Contagion appears to be excluded.

An alternative hypothesis is that the disease is due to a pathogen (e.g. a retrovirus or other type of mobile element) integrated in the genome. The element is assumed to be acquired either from an affected or predisposed parent or as a result of an early (and season-dependent) genetic re-arrangement which leaves the element in the same pathogenic form. If this "hot-spot" of genetic recombination has survival value in relation to cerebral growth mechanisms (e.g. by promoting the development of lateral asymmetry) this could account for the persistence of the disease in the face of a fertility disadvantage. The association between the "virogene" and the mechanisms of lateralised cerebral develop- ment provides a possible explanation for the predilection of the disease for the left hemisphere. It may also be relevant to timing of disease onset.

References

1. Torrey, EF and Peterson, MR (1976). The viral hypothesis of schizophrenia. Schizophrenia Bulletin, 2, 136
2. Crow, TJ (1978). Viral causes of psychiatric disease. Postgraduate Medical Journal, 54, 763
3. Hare, EH (1983). Was insanity on the increase? Brit.J.Psychiat. 142, 439
4. Torrey, EF (1980). Schizophrenia and Civilisation. Jason Aronson, New York.
5. Crow, TJ (1983). Is schizophrenia an infectious disease? Lancet, i, 173
6. Abe, K (1969). The morbidity rate and environmental influence in monozygotic co-twins of schizophrenics. Brit. J. Psychiat.115, 519
7. Crow, TJ and Done, DJ (1986). Age of onset of schizophrenia in siblings: a test of the contagion hypothesis. Psychiat.Res. 18, 107
8. Dalen, P (1975). Season of Birth: A Study of Schizophrenia and other Mental Disorders. Amsterdam: North Holland.
9. Crow, TJ (1984). A re-evaluation of the viral hypothesis: Is psychosis the result of retroviral integration at a site close to the cerebral dominance gene? Brit. J. Psychiat. 145, 243
10. Crow, TJ (1986). Left brain, retrotransposons and schizophrenia Brit. Med. J. 293, 3
11. Callahan, R, Chin, IM, Wong, JFH, Tronick, SR, Roe, BA, Aaronson, SA and Schlom, J (1985). A new class of endogenous human retroviral genomes. Science, 228, 1208
12. Steele, PE, Rabson, AB, Bryan, T and Martin, MA (1984). Distinctive termini characterise two families of human endogenous retroviral sequences. Science, 225, 943
13. Mager, DL and Henthorn, PS (1984). Identification of a retrovirus-like repetitive element in human DNA. Proc.Nat.Acad.Sci. USA. 81, 7510
14. Paulson, KE, Doka, N, Schmid, GW, Misra, R, Schindler, CW, Rush, MG, Kadyk, L and Leinwand, L (1985). A transposon like element in human DNA. Nature, 316, 359
15. Hickey, WE (1982). Selfish DNA: a sexually-transmitted nuclear parasite. Genetics, 101, 519
16. Crow, TJ (1986). The continuum of psychosis and its implication for the structure of the gene. Brit. J. Psychiat. (in press)
17. Jamison, KR (1986). Manic-depressive illness and accomplishment: creativity, leadership and social class. In: Goodwin, FK, and Jamison, KR (eds) Manic Depressive Illness. Oxford University Press.
18. Karlsson, JL (1984). Creative intelligence in relatives of mental patients. Hereditas, 100, 83
19. Noreik, K and Odegard, O (1966). Psychoses in Norwegians with a background of higher education. Brit. J. Psychiat. 112, 43
20. Colter, N, Crow, TJ, Frith, CD, Jagoe, JR, Johnstone, EC, Kreel, L and Owens, DGC. Developmental arrest of cerebral asymmetries in early onset schizophrenia (in preparation).

NEUROIMAGING

7
A neurodevelopmental perspective on brain pathology in schizophrenia
D.R. Weinberger

INTRODUCTION

Most neurobiological hypotheses about schizophrenia hold that the characteristic symptoms reflect dysfunction of the brain and that this dysfunction is the result of brain disease. It is generally assumed that there is a simple temporal relationship between the onset of the disease process and the clinical presentation of the illness. In other words, it is assumed that as the disease occurs, brain function is impaired and, _pari passu_, illness is manifest. This view would be consistent with the notion that schizophrenia is caused by a metabolic disorder or dysfunction of neural transmission, or with the notion that schizophrenia results from some episode of brain trauma such as might happen following viral encephalitis. It is also consistent with the possibility that schizophrenia is a progressive neurodegenerative disorder. To the extent that there is a genetic component to the illness, this may involve susceptibility to the disease process or the expression during early adult life of a genetically determined pathophysiology.

In each of these examples, it would be expected that if the etiological disease process also causes structural pathology of the brain, the development of this pathology would be temporally related to the development of the disease. In other words, if schizophrenia is caused by a viral infection in early adult life, then whatever pathology this infection produces, it should not be apparent before the infection occurs. If schizophrenia is analogous to a metabolic encephalopathy or to a neurodegenerative disorder, then the pathology should follow the disease process and become more extensive as the metabolic or degenerative process continues.

The results of recent neuropathological investigations of schizophrenia strongly suggest that the illness is associated with structural pathology of the brain, but that this pathology does not appear at the time of onset of characteristic symptoms and does not change in concert with changes in the clinical course [1]. If this pathology is related to the etiology of the illness, which seems likely, then the usual assumptions about the pathogenesis of this illness may need to be revised.

The following discussion will present one possible revision, a scheme for the pathogenesis of schizophrenia that is built around the interaction between a congenital brain lesion and normal brain development. It is based on several assumptions about the nature of the neuropathological findings in schizophrenia and on the clinical maxim that the manifestations of a brain lesion vary with the maturational state of the brain systems involved. After a presentation of the assumptions underlying this thesis, a synthesis of the implications of these assumptions for the pathogenesis of schizophrenia will be considered.

Assumption 1. <u>Schizophrenia is associated with structural pathology of the brain.</u> This assumption is based on the results of studies using the Computed Tomography or CT x-ray scan, Magnetic Resonance Imaging or MRI scan, as well as post mortem anatomical techniques. In the great majority of over 40 controlled studies with CT, patients with schizophrenia have been shown to have larger cerebral ventricles and more obvious cortical markings then normal individuals of the same age [2]. These findings have been described in first episode schizophreniform patients prior to the administration of psychiatric treatments [3,4], indicating that they are not the result of such treatments. Moreover, they are found primarily in the ill twin of dyscordant monozygotic twinships [5], suggesting that they are not artifacts of sociocultural circumstances or of psychosocial factors during childrearing.

The findings are subtle and require quantitative methods for them to be observed. They suggest that a pathological process has occurred in the brain and caused either a slight reduction in tissue mass or a failure of normal tissue development. The findings do not by themselves implicate a specific etiology, as they are, by definition, non-specific signs. There is some evidence that certain brain areas may be preferentially involved. Recent studies of the distribution of cortical markings have found that the differences between patients and normal controls are most marked in prefrontal cortex [2]. A similar finding has been reported in an MRI study [6]. Since MRI has greater potential as a technique for studying quantitative brain structure during life, additional information about the distribution of the CT findings will probably emerge from future studies with this technique.

Finally, as a result of the resurgence of interest in neuropathological studies of schizophrenia stemming from the CT findings, evidence for structural abnormalities has been reported in several recent postmortem studies [7]. While these studies have not provided a consistent picture of the pathology found at postmortem, in part because of different approaches to examining the tissue, they have suggested that pathology of the limbic system, especially amygdala and hippocampus [8-10], is a replicable phenomenon. As in the case of the CT findings, the results have been subtle and often have required morphometric, i.e., quantitative, techniques to be appreciated. It is worth noting that if the postmortem results can be replicated consistently, especially those implicating limbic pathology,then the basis for the CT findings will probably have been determined. Since the ventricles of the brain form much of the perimeter of the limbic system, pathology of the latter would be expected to produce enlargement of the former. The origin of the prefrontal changes observed with CT are less clear, although one recent postmortem study reported reduced neuronal cell counts in prefrontal cortex [11]. An alternative possibility is that the prefrontal changes are secondary to pathology in limbic sites which normally provide considerable input to prefrontal cortex.

Assumption 2. The structural brain pathology in schizophrenia is non-progressive. This assumption is based primarily on two experimental findings. First, lateral ventricular size on CT does not correlate with length of illness [2]. In the majority of CT studies, investigators have looked for a correlation between how long a patients has been ill and how much brain pathology exists. If the pathology were the result of an active neurodegenerative process or were progressive, such a correlation would be expected. Indeed, even the neuropathological changes associated with normal aging lead to increasing ventricular size over time. In the case of schizophrenia, the overwhelming majority of studies have failed to find a correlation between evidence of brain pathology on CT and length of illness.

The second experimental finding supporting this assumption are the results of a recent prospective follow-up study of CT scans in patients with chronic schizophrenia [12]. In this study, 15 patients had a repeat CT scan on the same model machine after seven to nine years following the initial scan. No changes in either ventricular size or cortical markings were found despite continuous hospitalization in half the cases and continuous neuroleptic treatment in all the cases. It is also worth noting that this assumption is consistent with the postmortem microscopic data. Most studies have not found evidence of an active neuropathological process, i.e., inflammation, proliferative gliosis, dying neurons, etc.

Assumption 3. The brain pathology in schizophrenia occurs early in development. There are several experimental findings that together lead to this assumption. First, ventricular enlargement has been found in first episode schizophreniform patients who are less than 20 years of age [8-10]. It is unlikely that ventricular enlargement would develop acutely in an adolescent without producing more traditional neurological symptoms. The more likely interpretation of this finding is that the pathology has been around for a while. The probability that it resulted from an early developmental neuropathological process is further suggested by the finding reported by several independent groups that ventricular enlargement in patients with schizophrenia is associated with a poor premorbid social history [2]. It is also supported by reports that ventricular enlargement is more common in patients who have a history (albeit retrospectively derived) of perinatal complications [13]. Finally, most of the postmortem studies have not found evidence of gliosis. Since gliosis is uncommon when pathology occurs before one or two years of postnatal life, this observation also suggests that the pathology occurred early in development.

Assumption 4.   The clinical manifestations of brain pathology vary depending on the ontological maturity of the areas involved. This assumption is based on clinical experience in man and on basic experiments in sub-human primates that demonstrate that the behavioral implications of even a fixed structural brain lesion change as the subject ages. A clear illustration of this is the changing pattern of clinical symptoms and signs seen in cases of cerebral palsy, a condition caused by perinatal trauma and/or fetal developmental disorders. While the brain pathology in cerebral palsy is fixed and does not change after a few months of postnatal life, the clinical manifestations of the pathology change with time. Pathology that presents with hypotonia and delayed motor milestones at one year of age may be manifest as a hemiparesis at two and cause athetosis to appear at four. Seizures may present at seven years of age and the pattern of the seizures may not include psychic symptoms until the individual reaches adolescence or early adult life. This changing clinical story in the context of a fixed congenital lesion is thought to reflect an interaction between the lesion and maturational changes that occur in areas involved by the pathology. In other words, unless the motor systems that mediate athetosis are functionally mature, the fact that these systems are structurally compromised may be clinically inapparent until they have reached the necessary stage of ontological development.

Another illustration of this assumption is the relationship between the time of onset of certain neurological conditions that occasionally present with psychopathology and the nature of the psychopathology. Patients with Huntington's Disease, Wilson's Disease, CNS Lupus, head trauma, brain tumor, etc, may present with psychosis, or sometimes with depression, or possibly dementia. A review of the clinical literature suggests that the critical determinant of whether one of these conditions presents with schizophreniform psychosis or with other psychiatric symptoms is the age of the patients [14]. It is much more likely that if psychosis is the presenting symptoms that the individual will be in late adolescence or in early adult life then that he or she will be older. This suggests that there is something "generic" about early adult life and the expression of psychotic symptoms regardless of the brain pathology responsible. Using the cerebral palsy analogy, it suggests that a lesion in brain systems that mediate psychotic behavior may not be manifest as psychosis until this critical period.

Assumption 5.   The pathology in schizophrenia affects the function of brain systems that mature relatively late in development. Although the specific neural systems responsible for schizophreniform behavior are unknown, there is considerable circumstantial evidence to suggest that limbic and cortical dopaminergic pathways are involved in the expression of so-called positive psychotic phenomena and prefrontal cortex is involved in the characteristic defect symptoms and intellectual impairment [14]. Recent postmortem studies [15] and in vivo cerebral imaging of dopamine receptor activity [16] suggest that both cortical and subcortical dopamine systems reach peak activity in early adult life around the time of onset of schizophrenia. In fact, these studies suggest that dopaminergic activity declines fairly rapidly during middle age so that by age 40, there is virtually one-half the level of activity seen at age 20. In women, dopaminergic activity appears to peak approximately five years later and to decline less precipitously. This latter finding may have implications for epidemiological data that suggested a later age of onset and a less malignant course of schizophrenia in women. It follows from these observations that if the brain pathology in schizophrenia affects the mechanisms that normally modulate dopaminergic function, the expression of dysfunction, especially if this dysfunction results in relative overactivity of dopaminergic neural transmission, is likely to become most apparent in early adult life.

In the case of the prefrontal cortex, recent data suggest that this area shows signs on CT and MRI of structural pathology (vide supra) and on the basis of physiological cerebral imaging (i.e., PET, rCBF) of physiological pathology [17]. The prefrontal cortex especially its dorsolateral aspect, is possibly the last brain region to reach anatomical and physiological maturity, Yakovlev [18]. Studies in subhuman primates and in man indicate that the myelogenic cycle for this area is late, occurring in man primarily from the late second to the middle third decades of life. The cognitive and behavioral functions subserved by prefrontal cortex (e.g., Piagetian "formal operations") also are not firmly established until late adolescence or early adulthood. By analogy to the case of cerebral palsy, it might be expected that pathology of prefrontal cortex, especially subtle pathology, would be relatively inapparent until early adult life. This exact scenario has been described in the monkey. Goldman [19] showed that a perinatal fixed lesion of the dorsolateral prefrontal cortex of the rhesus monkey did not significantly impair the cognitive behavior of the preadolescent monkey; yet, when the same monkey reached sexual maturity, it began to manifest major intellectual and behavior deficits.

SYNTHESIS

The assumptions detailed above about the brain pathology in schizophrenia and its relationship to certain highly evolved brain systems that normally mature late in the ontological scheme of brain development suggest a novel approach to conceptualizing the pathogenesis of this disorder. The approach is anetiological and would be consistent with any cause that could be responsible for early developmental pathology of limbic system and prefrontal cortex. It is not improbable that multiple causes for the brain pathology in schizophrenia exist. Nevertheless, it is also probable that a final common pathophysiological pathway translates the possibly diverse etiologies into a somewhat consistent clinical entity.

The final common pathophysiology proposed here involves a fixed congenital lesion and an age-dependent and changing state of normal brain function. The lesion is nonspecific and affects interrelated limbic-diencephalic nuclei and prefrontal cortex. The onset of schizophrenia does not appear to be related to the time of occurrence of the lesion, which seems to be congenital. Instead, the onset of this illness may be linked to the time at which the brain areas involved by the lesion, particularly dorsolateral prefrontal cortex, reach functional maturity and to when brain dopaminergic activity peaks. In light of recent research in laboratory animals that suggests a role of prefrontal cortex in modulating limbic dopamine activity [20], it is possible that a lesion similar to the one implicated in schizophrenia, by impairing prefrontal function, could also result in a dysregulation of limbic dopaminergic activity. Because late adolescence is a critical maturational time for both prefrontal neural systems and for limbic dopamine systems, a lesion that effects the function of these interrelated networks might have its maximum clinical impact at this time of life. Finally, the course of the illness, in particular the lessening of psychotic or "positive" symptoms and the increase in defect symptoms over time, parallels the normal reduction of brain dopaminergic activity that occurs with aging.

REFERENCES

1. Weinberger, DR (1984). Computed tomography (CT) findings in schizophrenia: Speculation on the meaning of it all. J. Psychiatr. Res., 477-490
2. Shelton, R and Weinberger, DR (1986). CT scan studies of schizophrenia. In: Nasrallah, HA and Weinberger, DR (eds.) "The Neurology of Schizophrenia". p. 207-250. (Amsterdam: Elsevier N Holland)
3. Weinberger, DR, DeLisi, LE, Perman, G, Targum, S and Wyatt, RJ (1982). Computed tomography scans in schizophreniform disorder and other acute psychiatric patients. Arch. Gen. Psychiatry, 39, 778-783
4. Nyback, H, Berggren, BM and Hindmarsh T (1982). Computed tomography of the brain in patients with acute psychosis and in healthy volunteers. Acta. Psychiatr. Scand, 65, 403-414
5. Reveley, AM, Reveley, MA, Clifford, CA and Murray RM (1982). Cerebral ventricular size in twins discordant for schizophrenia. Lancet, i, 540-541
6. Andreasen, N, Nasrallah, HA, Doran, V et al. (1986). Structural abnormalities in the frontal system in schizophrenia. Arch. Gen. Psychiatry, 43, 157-160
7. Kirch, D and Weinberger, DR (1986). Post-mortem histopathological findings in schizophrenia. In: Nasrallah, HA and Weinberger, DR (eds.) "The Neurology of Schizophrenia". p. 325-348. (Amsterdam: Elsevier N Holland)
8. Bogerts, B, Meertz, E and Schonfeldt-Bausch R (1985). Basal ganglia and limbic system pathology in schizophrenia: A morphometric study. Arch. Gen. Psychiatry, 42, 784-791
9. Brown, R, Colter, N, Corsellis, JAN, Crow, TJ, Frith, CD, Jagoe, R, Johnstone, EC and Marsh, L (1986). Postmortem evidence of structural brain changes in schizophrenia. Arch. Gen. Psychiatry, 43, 36-42
10. Kovelman, JA and Scheibel AB (1984). A neurohistological correlate of schizophrenia. Biol. Psychiatry, 19, 1601-1621

11. Benes, FM, Davidson, J and Bird ED (1986). Quantitative cytoarchitectural studies of the cerebral cortex of schizophrenics. Arch. Gen. Psychiatry, 43, 31–35

12. Illowsky, B, Juliano, D, Bigelow, LB and Weinberger DR. Stability of CT findings in schizophrenia: An eight year follow-up study (submitted)

13. Williams, AO, Reveley, MA, Kolakowska, T, Andern, M and Mandelbrote BM (1985). Schizophrenia with good and poor outcome. II. Cerebral ventricular size and its clinical significance. Br. J. Psychiatry, 146, 239–246

14. Weinberger DR. Implications of normal brain development for the pathogenesis of schizophrenia. Arch. Gen. Psychiatry, (in press)

15. Bzowej, NH and Seeman P (1985). Age and dopamine D2 receptors in human brain. Soc. Neurosci. Abstr, 11, 889

16. Wong, DR, Wagner, HN Jr, Dannals, RF, Links, JM, Frost, JJ, Ravert, HT, Wilson, AA and Rosenbaum AE (1984). Effects of age on dopamine and serotonin receptors measured by positron emission tomography in the living human brain. Science, 226, 1393–1396

17. Weinberger, DR and Kleinman JE (1986). Observations on the brain in schizophrenia. In: Hales, RE and Frances, AJ (eds.) "Psychiatry Update, American Psychiatric Association Annual Review, Vol. 5". pp. 42–67. (Washington: APA Press)

18. Yakovlev, PI and LeCours, A-R (1964). The myelogenetic cycles of regional maturation of the brain. In: Minkowski, A (ed.) "Regional Development of the Brain in Early Life". pp. 3–70. (Oxford, England: Blackwell Scientific Publications)

19. Goldman, PS (1971). Functional development of the prefrontal cortex in early life and the problem of neuronal plasticity. Exp. Neurol, 32, 366–387

20. Pycock, CJ, Kerwin, RW and Carter, CJ (191980). Effect of lesion of cortical dopamine terminals on subcortical dopamine in rats. Nature, 286, 74–77

8

Neuromorphological correlates of schizophrenic disorders: focus on cerebral ventricular enlargement

E. Sacchetti, A. Vita, A. Calzeroni, G. Invernizzi and C.L. Cazzullo

---

INTRODUCTION

The search for some visible basis for insanity is certainly not recent. We know that even in the Middle Ages the allegory of the "madness stone" had stable place in the popular thought of the time.

Only in this century, however, has the neuromorphological approach to psychiatric disorders been the focus of explicit scientific attempts at verification. Macroscopic post-mortem investigations (for review see 1) and especially "in vivo" studies employing pneumoencephalography (PEG) (2,3,4) represent the first and very notable experimental efforts in this area.

In spite of the relevance of these "advances" and their possibilities for further development, the neuromorphological approach was progressively laid aside during the 60's. Several factors contributed to this tendency, including, for example, the risk of dealing with artifacts and the difficulty of finding valid control groups. For the post-mortem investigations there were the inevitable problems of retropsective correlations with clinical data and locating populations suitable for examination. PEG studies met up with the problem of the invasiveness and potential danger of the method.

In the recent past, the advent of new cerebral visualization techniques -computed tomography (CT), positron emission tomography (PET) and magnetic resonance imaging (MRI)- opened a new and very promising era for the "in vivo" morphological approach to psychiatric disorders. In addition to permitting us to obtain detailed anatomical data and, in the case of PET scan, previously precluded functional analyses of specific cerebral areas, these techniques have enabled us to overcome many of the limitations to which early pioneering PEG studies were subject. Consequently, the past few years have brought us accumulating findings on the brain imaging correlates of various psychiatric illnesses such as affective disorders (5,6), anorexia (7), childhood autism (8) and Alzheimer's disease (9).

Thanks to these new techniques our knowledge of schizophrenia, in particular, has advanced both qualitatively and quantitatively. The fact that schizophrenia has been the priveleged investigative area in which these new technologies have been applied is not surprising when we consider that the classic PEG investigations, showing clear signs of cerebral atrophy in many patients, represented a point of departure for studies not limited to exploration alone.

After 10 years of intensive CT research subsequent to Johnstone et al's pioneering first findings (10), it is possible to firmly conclude that in many instances schizophrenia conforms to organic hypotheses and that the evidence for these raises a legitimate doubt about the term functional psychosis and whether it can any longer be validly applied to the disease.

Dilated cortical sulci, atrophy of the anterior cerebellar vermis, reversal of the frontal and occipital lobe asymmetries, reduced brain density have all been reported (for reviews see 11,12). More frequent than any others, however, are reports on a pathological enlargement of the lateral cerebral ventricles, demonstrable in about a third of the cases.

This observation inevitably leads to a number of questions requiring specific answers.

First, are enlarged lateral ventricles a significant or irrelevant characteristic? Do they occur by chance or because of an inadequate control of variables extraneous to the disorder? If enlarged ventricles are significant, are we dealing with a pathogenetically salient abnormality? In other words, is there an actual link between the causal mechanisms of ventricular enlargement and those of schizophrenia? On the other hand, are we merely dealing with a simple marker of a schizophrenic subtype? Aside from whether enlarged ventricles represent a simple marker or a pathogenetic abnormality, do patients with the abnormality consitute a subgroup with other distinctive features? This subgroup of patients can satisfactorily fall into any of the subtypes of schizophrenia so far proposed in nosography? But should this not be the case, does the subgroup have the heuristic value and clinical significance that could make a change in our diagnostic procedures advisable? Moreover, is ventricular enlargment connected with a predisposition to schizophrenia or rather with the disease itself? If ventricular dilatation is not a real index of vulnerability, does it appear with the onset of the first symptoms or is it an initial or terminal phenomenon of the natural or pharmacologically-influenced evolution of the disorder?

The data from several laboratories are most consistent and no longer throw any doubt on the question of whether the large number of schizophrenic patients with ventricular enlargement is due to chance. These findings certainly speak in favour of Ventricular Brain Ratio (VBR) as a supplementary tool valuable for assisting clinicians and researchers to subclassify schizophrenics.

It is also certain that many of the questions raised still need rigorous investigation. Sometimes the conclusions from different studies conflict. Even when results are not apparently controversial, it is evident that they come from studies conducted with samples that are too small and/or have not been verified in adequate replication analyses. Further, patient recruitment often is not randomized but merely "ad hoc" on the basis of specific pecularities inherent in the variable chosen for investigation.

Under these conditions what appears questionable is the generalizability of the conclusions and therefore the possibility of using the results of the various studies in a uniform interpretative context.

Further, the experimental designs implemented do not even seem to safeguard against the risk of spurious associations due to various sources unrelated to ventricular enlargement but affecting other parameters under examination.

Given the above considerations discussed, we believed it necessary to further examine the existing knowledge of ventricular enlargement in schizophrenia.

Over the course of the past few years we have been carrying out a large-scale study of the association between ventricular size and several clinical and biological variables relevant to schizophrenia research (6,13,14). This chapter summarizes the findings from these studies, some of which are still underway.

SUBJECTS AND METHODS

Patients.

Two-hundred-seventeen consenting patients underwent noncontrast CT scans, having fulfilled the following inclusion criteria: 1. Diagnosis of schizophrenia, schizophreniform disorder, schizotypal or paranoid personality disorder in accordance with DSM-III criteria (15); 2. Less than 45 years old; 3. No evidence of existing medical or neurological illnesses that could be responsible for cerebral atrophy or psychiatric symptoms as established by physical examination and instrumental and laboratory analyses; 4. No history of alcohol or drug abuse, head trauma with loss of consciousness or seizure disorders; 5. No intake of steroid medication in the 3 months prior to the CT scans.

There were 189 schizophrenic patients (127 males, 62 females, mean age 26.4 ± 6.9 years), 17 with schizophreniform disorder (13 men, 4 women; mean age 23.5 ± 6.9 years) and 11 with personality disorders (7 schizotypal, 4 paranoid; mean age 29.2 ± 13 years).

According to the DSM-III criteria for subtyping schizophrenia, there were 98 patients with disorganized, 53 with undifferentiated, 32 with paranoid, 2 with catatonic and 3 with residual schizophrenia. Fifty-nine patients had subchronic, and 130 chronic schizophrenia.

Control group.

The control group consisted of subjects who had CT scans as part of routine diagnostic examinations to exclude brain injury from minor accidental head trauma without loss of consciousness. Subjects were excluded if they had symptoms or signs of neurological damage or if their CT scans showed accident-related lesions. The criteria 2 to 5 for patients selection and the absence of a past history or present evidence of psychiatric disorder were also requirements for inclusion. Fifty-seven subjects (33 men, 24 women; mean age 25.4 ± 8 years), formed the control group.

Methods

CT scans were performed with a model 1010 EMI scanner and, in a few cases (30 schizophrenics, 11 personality disorders), with a 9000 II GE scanne... Ten to 12 slices were obtained for each patient and control subject.

Cerebral ventricular size was measured on the CT slice showing the lateral ventricles at their largest by means of a manual planimetric grid method similar to that described by Benes et al. (16). Graph paper with millimeter intervals was used as a grid for tracing the perimeter of the brain and of each ventricle. A program for automatic computing of the cerebral and ventricular areas was devised on the basis of the numerical input of the readings of the brain and ventricular edges at each millimeter interval. The sum of the lateral ventricle areas was then divided by the brain area and the result multiplied by 100: the resulting number was the VBR (17). This method for determining ventricular size was used because it is very easy to employ and proved to be as reliable as evalutation by mechanical planimetry. Moreover, it gave the same results as those obtained by the mechanical planimetric method and is therefore likely to have equally good correlations with ventricular volume (18).

All the areas were traced twice by two independent raters who were unaware of the subject's diagnosis. The mean of the four resulting ratings was used.

Inter-rater reliability was measured and found to be high (r=.95), with a mean difference of .5 VBR units between the two raters.

Additional criteria and procedures for the independent studies were as follows: a) Age at onset of schizophrenia was defined, when possible, as the patient's first psychiatric referral (hospitalization or treatment in an outpatient facility) or the period during which the patients' relatives or friends noticed marked changes in his behavior and social functioning; b) The clinical pictures of schizophrenic onset patients who had never received psychopharmacological drugs were analyzed and the following symptoms rated as present or absent: delusions, hallucinations, positive formal thought disorders, social withdrawal, poverty of speech, bizzarre behavior, affective disturbances; c) Neuropsychological evaluations were obtained by administration of the Wechsler Adult Intelligence Scale (WAIS) and the Standardized Luria Nebraska Neuropsychological Battery (LNNB) (19). LNNB raw scores, T-scores and the differences between T-scores and patient's critical level (a measure which takes patient age and education into account) were considered; d) The clinical outcome of neuroleptic treatment was evaluated in patients who had been hospitalized for an acute psychotic episode while off drugs. All were assigned to a standard neuroleptic therapy and entered a 1 year follow-up during which clinical outcome was evaluated monthly by means of the Brief Psychiatric Rating Scale (BPRS) (20); e) Data on family history (FH) of schizophrenia was collected by means of separate interviews with each schizophrenic patient and at least one of his relatives. Positive FH patients were defined as those with at least one first degree schizophrenic relative and negative FHs those who had no affected first or second degree relatives. Patients with only one second degree schizophrenic relative and those for whom information was unreliable or lacking were excluded from study; f) HLA-A, -B, and -C typing was carried out according to the lymphocytotoxic microtechnique (21). HLA-DR antigens were determined by use of UCLA DRw typing trays. The antigens frequencies of patients were compared to those of 150 age and sex-matched healthy adult blood donors; g) Smooth pursuit eye movements (SPEMs) were assessed on a pendulum eye-tracking task and recorded with an electrooculographic technique. SPEMs were then rated on a qualitative 5-point scale and with a quantitative index expressing the number of eye movement arrests/cycle (22).

Statistical analyses included Pearson's correlation coefficient, paired and unpaired Student's t tests, analysis of variance, analysis of covariance, Chi square, Friedmann's X^2 test.

RESULTS

The mean VBR of our 189 schizophrenic patients was 5.3 $\pm$ 3, significantly higher than that of the 57 age- and sex-matched healthy controls (3.5 $\pm$ 2; t= 4.8, p<.0001). (table 1)

Table 1. Demographic characteristics and VBR values for patients with schizophrenic disorder and for controls

		Healthy controls	Schizophrenic disorder
N.		57	189
AGE (years)	-mean $\pm$ SD	24.5 $\pm$ 8	26.4 $\pm$ 6.9
	-range	13 - 45	14 - 45
SEX (M:F)		33 : 24	127 : 62
VBR	-mean $\pm$ SD	3.5 $\pm$ 2	5.3 $\pm$ 3*
	-range	.5 - 9.6	.8 - 15.4
VENTRICULAR ENLARGEMENT: N.(%)		3 (5)	44 (23)

* Student's t test = 4.8, p<.0001

Cerebral ventricular enlargement (VE) was defined as a VBR exceeding the mean of controls by more than 2 SD, in accordance with the large majority of the authors. The threshold for enlargement, therefore, resulted of 7.5 VBR units in the present study, a value slightly lower than that reported in other studies which, however, included controls in a wider age-range, with

71

subjects over 45 years also represented.

According to this criterion, 44 of our patient, i.e. 23,3% of the entire patients sample, had ventricular enlargement (table 1). This is an intermediate figure among those reported in the literature to date (6 to 60%) and we believe it represents an adequate estimate of the prevalence of enlargement in schizophrenia for the large size and random composition of our sample.

The question then arose of whether patients with and without VE meaningfully differed in some relevant clinical or biological aspects.

Ventricular enlargement and the temporal course of schizophrenia.
An adequate definition of the temporal relationship between the appearance of VE and that of schizophrenic disorder obviously is basic for ascertaining whether we are dealing with a putative index for distinguishing meaningful subgroups of patients, or only with the end product of the evolution of the disease or the events inextricably connected with it.

On the whole, tha data available indicate that VE is not simply the consequence of processes related to the temporal course of schizophrenia. With only a few exceptions, the consensus is that VBR values do not correlate with the duration of illness (11,12); neither does patient age appear to have an influence, at least for those under 40 (11,12). This gives further, but indirect, support to the fact that VE is not mainly an "endowment" of cases in a temporally advanced stage of illness.

Further, there is some experimental evidence for the hypothesis that VE even predates the onset of schizophrenia. The isolated reports of abnormal VBR values in patients with schizophreniform disorders (13,23,24), or with personality disorders of the schizophrenic spectrum (6,25) as well as the evidence that patients with VE have increased chances of having had obstetric complications (26,27) and, possibly, a specific birth season (28) all speak in favour of this possibility.

Nonetheless, VE prevalence rates sometimes have been reported to increase with increasing age and with duration of illness (29,30).

In view of the above, it seemed worthwhile to newly examine the problem of time of appearance of VE in schizophrenic patients, combining cross-sectional and longitudinal data.

For the 157 schizophrenic patients tested so far, VBR values were not correlated with duration of illness (r=.08, p=NS), nor with the age of the patients (r=.03, p=NS). No correlation was found with the third temporal variable considered, age at onset (r=.02, p=NS).

Other findings further corraborate these results.

First, the dichotomization of patients into the DSM III subchronic and chronic subgroups was not able to explain inter-patients differences in cerebral ventricular size. The 59 subchronic and 130 chronic patients overlapped both relative to absolute VBR values (t=.06, p=NS) and the incidence of cases with VE (n=31, i.e. 24% and respectively n=13, i.e. 22%; X^2 =.4, p=NS).

Equally irrelevant was the effect of the treatment status.In fact, first admission schizophrenics never treated and those treated with low doses (doses equivalent to not more than 3 mg of haloperidol/die) for not more than 2 months had VBR and VE incidence values identical to those in the entire patient population.

Furthermore, when 9 patients were rescanned 18 to 61 months after their first examination, their VBR values had not significantly changed (paired t test= .23, p=NS) (figure 1)

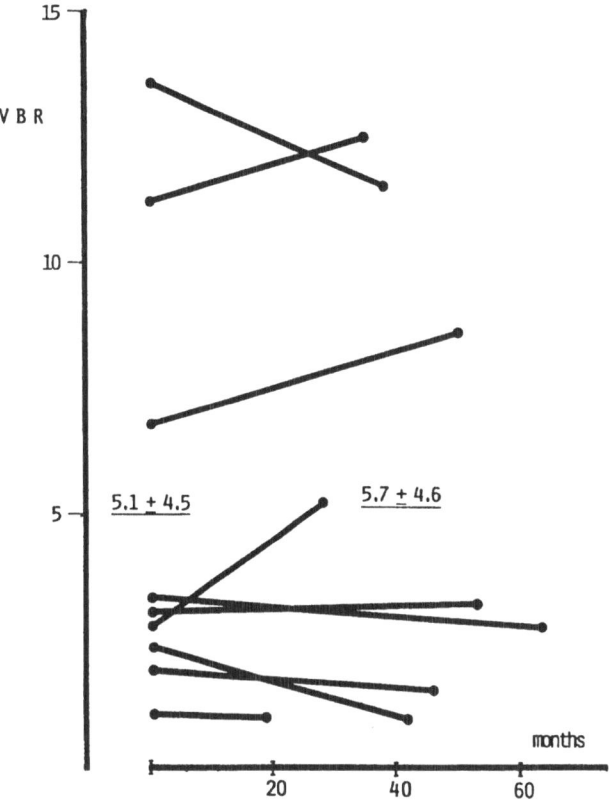

FIGURE 1   CT scan follow up study of schizophrenic patients. No significant change of VBR values was found when patients were rescanned after 18 to 61 months since the first examination (mean time interval between scans: 39.4 + 14.3 months). The correlation coefficient between the first and follow-up VBR values was .94.

Finally, 17 schizophreniform disorder and 11 personality disorder patients had VBR values (6.6 $\pm$ 2.9 and 4.7 $\pm$ 3.1) significantly higher than those found in healthy controls (t= 5.2, p<.001 and t= 2.1, p=.03) and equal to those found in schizophrenic patients (p=NS in each case). They also had an incidence of pathological ventricular dilatation similar to that found in the schizophrenics (35% and 27% respectively).

Interestingly, all the patients with schizophreniform disorder, whether outpatients or first-admissions to the Psychiatric Clinic, were scanned before starting with antipsychotic medication. All the patients with personality disturbances were outpatients who had never been hospitalized and, except for a few who had previously taken benzodiazepines or antidepressants sporadically, the group was negative for previous therapy with psychotropic drugs.

All these findings shed more light on the question of the time that VE appears in schizophrenic patients and of the effects of time-related variables on this neuromorphological index.

First, since VBR did not correlate with duration of illness and the data on subchronic patients were no different from those on chronic ones, there is thus evidence for the fact that VE is a result of processes already in play from the earliest stages of the disorder. Even if definite conclusions can be drawn only after study of the clinical outcome of these patients, the acquired evidence that ventricular dilatation is also found in patients with schizophreniform or personality disorder (i.e. in patients who belong to groups at high risk for developing schizophrenia) appears to suggest that the time of the appearance of this abnormality in schizophrenic patients probably occurs prior to the onset of the disorder, perhaps even a great deal of time before.

This would imply that many of the conclusions above VE in schizophrenia should be placed in a broader general context of a continuous spectrum of disturbances ranging from schizophreniform and some personality disorders to the most severe and relatively late limit of schizophrenia.

Further, since VBR was stable over time and did not correlate with the duration of the illness, it is strengthened the hypothesis that factors related to the evolution of the illness, such as length of hospitalization, amount and duration of neuroleptic treatment, and the many consequences of the progressively poor social adaptation which is typical of schizophrenia do not play either a primary or primitive role on VE.

However, this conclusion does not exclude the possibility that long term exposure to the various factors inherent in the disease course processes may have a further effect on dilatation and/or that the chance for this supplementary dilatation may affect the patients in different ways according to whether they belong to the group with enlarged or normal ventricles. These hypotheses, naturally, should be tested by means of long-term follow-up studies.

Diagnostic subtypes and symptoms
Dating from Kraepelin (31) right up to the present, rigorous descriptive efforts for subtyping schizophrenia by means of symptomatological characteristics putatively assumed as distinctive

undoubtedly have informed a large part of the most relevant approaches for diagnosing this disorder. This classificatory approach has been carried forward without regard to the fact that schizophrenia was placed in an etiological framework, as a simple disease entity, a group of related diseases or a group of separate diseases.

From this it appears comprehensible that one of the major goals of research on cerebral ventricular size in schizophrenia was verification of the possibility that VE occured in patients who were of different clinical subtypes. Research in this area has mainly involved the classical conventional diagnostic subtypes and the dichotomy positive and negative symptoms because of its important role in clinical practice and theory.

As far as the classical diagnostic subtypes are concerned, the evidence accumulated are intriguing. In some instances (32,33) no association has been found with VE, while in other cases some association has been claimed (34,35,36). Further, analysis of the results that have found some association clearly indicates that great discrepancies do exist, VE now concentrated among paranoids rather than hebephrenics (34), then among paranoids and hebephrenics as compared to undifferentiated patients (35) or among residual schizophrenics rather than paranoids (36).

With regard to the positive and negative symptoms dichotomy, a more definite trend seems to emerge. Johnstone et al's initial report (10) of an association between VE and negative symptoms was followed by other independent ones that supported a predominance of negative symptoms and, concomitantly, a degree of rarity of positive symptoms among schizophrenics with ventricular dilatation (37,38). However, there are important studies (39,40) that maintain that VBR and both types of symptoms are independent or even that there is an increased frequency of some defined negative and positive symptoms among patients with absolutely normal VBR values (41).

Both in the case of studies mainly involving diagnostic subtypes and those concentrated on symptoms, the discrepancies among the results may undoubtedly be partiallty due to the fact that many studies have dealt with samples too small to be considered truly representative and that subtyping criteria among them were not always explicit or corresponding.

Further, the results sometimes refer to patients with associated features that are overtly atypical and not at all similar to those habitually associated to a defined subgroup of patients. An example of this is that when an excessive number of cases of VE was found among paranoids, the latter also were classified as neuroleptic non responders (34, 35).

But a particularly relevant confounding factor is that, for the most part, patients have been grouped irrespectively to the evolutional stage of the disorder, to the treatment status, and to the severity of the disease course. This has occured despite the fact that all these variables interacting in very complex ways undoubtedly have an influence on the patient's clinical characteristics. Let us take the case of negative symptoms. These not only represent a possible consequence of social and institutional reactions on positive symptoms occuring for extended periods, but also may be relatively overrepresented as compared to positive symptoms in patients drug treated for an acute decompensation both because they do not respond to neuroleptics as well and because such symptoms may be partially induced by the pharmacological therapy itself.

In order to partially overcome these limitations we considered it worthwhile to promote two different approaches.

The first approach is to specifically determine whether VE is preferentially confined within some defined diagnostic subgroups of schizophrenia. This is based on the assumption that a sufficiently sizeable sample selected on a rigidly randomized basis within a naturalistic setting could guarantee a partial but sufficient balance of several sources of variation.

The second approach, specifically addressed to learning more about the nature of the relationship between VBR and symptoms was, instead, based on a careful selection of schizophrenic patients at the onset of their disorders who never received neuroleptic therapy.

With respect to the diagnostic subtypes, we compared the ventricular size of 98 disorganized schizophrenics with that of 53 undifferentiated and 33 paranoid schizophrenics and found that the 3 subgroups were identical both in terms of absolute VBR values (5.3 ± 3.0, 5.4 ± 3.3 and 5.2 ± 3.3, respectively; Anova: $F(2,181)=.06$, p=NS), and the incidence of cases with enlarged ventricles (23, 13 and 7 subjects respectively, i.e. 23.4%, 24.5% and 21.2%; X^2 (d.f. 2)= .08, p= NS) (table 2).

Table 2. Characteristics and VBR values of patients with different diagnostic subtypes of schizophrenia (DSM-III criteria)

	DIAGNOSTIC SUBTYPES		
	DISORGANIZED	PARANOID	UNDIFFERENTIATED
N.*	98	33	53
AGE** (years $\pm$ SD)	25.2 ± 5.3	30.5 ± 7	25.3 ± 7.2
SEX (M:F)	65 : 33	23 : 10	38 : 15
VBR (mean $\pm$ SD)	5.3 ± 3	5.2 ± 3.3	5.4 ± 2.8
VENTRICULAR ENLARGEMENT: n.(%)	23 (23)	7 (21)	13 (24)

** F (2,181)= 9.1, p<001
* Two patients with catatonic and 3 with residual schizophrenia were not included in the table and in the analysis.

As far as the symptoms at onset were concerned, among the 20 naive schizophrenic patients that we selected, we found that patients with (n= 7) and without VE (n= 13) had equal relative risks for positive formal thought disorders; patients with VE had only a slightly increased risk for hallucinations (1.55) and alogia (1.7), but had almost three times the risk for affective disturbances such as affective flattening and almost half (0.45) for delusions (table 3).

The lack of any association between VE and some definite subtypes of schizophrenia must be considered as substantially conclusive. The size of the sample tested and the absolute identity of VBR values among the groups of paranoids, undifferentiated and

Table 3. Relative risks of different symptoms at the onset of schizophrenia in patients with ventricular enlargement respect to patients with normal ventricles.

SYMPTOM	Relative Risk
"Positive" symptoms	
-Delusions	.45
-Hallucinations	1.55
-Positive formal thought disorders	.875
-Bizzare behavior	1.2
"Negative" symptoms	
-Affective disturbances	3.0
-Alogia	1.69
-Social withdrawal	.75

disorganized patients imply that some kind of inversion of this tendency can be obtained only by testing a very large sample of patients with morphological and clinical characteristics completely different from those found in the sample discussed here. Since patients were selected on a purely randomized basis it seems plausible to maintain that the above would be hardly like to occur.

On the contrary, findings on VE and symptoms in naive schizophrenics must be considered parsimoniously in that they are decidedly preliminary, because of the limited size of the sample from which they come.

Furthermore, it should be noted that the data on diagnostic subgroups and those on symptoms are not rigorously comparable. The first refer to patients checked at various points of time during the course of their disorder while the second relate to patients checked in a similar initial stage of their disorder. In the absence of longitudinal studies specifically addressed to the verification of this possibility, we cannot exclude that patients with and without VE initially have non identical clinical characteristics but they progressively arrive at overlapping phenomenological presentations because of the effect of partially distinctive features of the natural or therapeutically-induced disease evolution.

Apart from these considerations, it should therefore not be surprising that on the one hand we saw that VE cut substantially across the diagnostic subtypes of schizophrenia and, on the other, that the patients had different symptoms in function of their VBR values.

Even though symptoms represent the guide-line criteria for defining diagnostic subgroups, the symptomatic approach and the syndrome approach are not really interchangeable. Differences in certain symptoms do not in themselves imply different subtypes of schizophrenia and, conversely, different subtypes of schizophrenia often present extensive similarities in psychopathological manifestations.

Finally, if VE effectively represents an acceptable marker for enucleating a homogeneous subgroup of schizophrenic patients with special etiopathogenic features, then the strict similarity of VBR values encountered among disorganized, undifferentiated and paranoid patients represents a further limitation for considering clinical subtypes of schizophrenia not as mere cross-sectional descriptions but as stable clinical entities.

Intellectual and neuropsychological functioning

As a consequence of the first CT study (10) which reported a significant relationship between lateral VE and poor cognitive performance in schizophrenia, impairment of intellectual and neuropsychological performance has been one of the major characteristics suspected to be associated with structural brain abnormalities in schizophrenia and this was immediately included among the relevant features of the "organic" type II schizophrenic syndrome (42).

Since that time, the literature on the relationship between higher nervous activity functions and structural brain abnormalities has swiftly accumulated (for review see 11,12) with great discrepancies in results.

A series of different test batteries was employed to explore cognitive and intellectual functioning in schizophrenic patients (Withers and Hinton, Folstein Mini-Mental State, Halstead Reitan Neuropsychological Battery, Luria Nebraska Neuropsychological Battery, WAIS, and Strub and Black's Mental Status Examination), used in relation to both ventricular size and degree of cortical atrophy. Not one of pairs of studies using the same battery, however, ever yielded the same results (42).

On the whole, schizophrenic patients with brain atrophy seem to perform poorly on neuropsychological tests, but the contradictory findings indicate the need for further study in this area. Further, these studies have not taken variables such as intellectual functioning and educational level into systematic account. These variables represent first-order issues in schizophrenia, since they have potential for baising neuropsychological test results (43).

This situation points to the need to undertake further research on the problem, with studies experimentally designed to take possible intervening variables into account. Consequently, we reconsidered the existing relationship between ventricular size and intellectual and neuropsychological performance in schizophrenia by administering the standardized LNNB and WAIS to 32 schizophrenic patients (mean age 25.1 ± 3.8 years; 22 males, 10 females) (44).

In our sample, VBR was inversely correlated with the patients' verbal IQ ($p=.04$), but not with performance or total IQs or any of the WAIS subscales.

On the other hand, there were significant positive correlations between VBR and 2 of the 14 LNNB scales, the T tactile scale ($p=.017$) and the T writing scale ($p=.046$). After transforming T to D scores (T score - critical level), however, the only significant correlation was that between VBR and the LNNB tactile scale($p=.024$).

Surprisingly, we did find significant correlations between 13 of the 14 LNNB T scores and both verbal and performance IQs, the only scale not related to the IQs being the tactile one. Further, eleven T scales were inversely correlated with patients' educational level, expressed in terms of years of school attendance. Given these results, we reanalyzed the relationship between VBR values and LNNB scores, taking into account the patients' IQs and educational levels. Partial correlation analysis showed that the correlation between both T writing and T tactile scores and VBR no longer was significant when verbal and performance IQs or educational level were controlled. Conversely, the correlation between D tactile scores and VBR values remained significant after undertaking all control procedures.

These results do not support previous findings (by Golden et al. (45)) on a definite relationship between brain atrophy and degree of LNNB deficits in young schizophrenic patients. Since Golden's patients had much higher VBR values and a greater frequency of dilatation than those included in our study, it is conceivable that correlations between ventricular size and neuropsychological deficits could particularly apply to patients with extremely high VBR values.

In any case, the present finding of a weak association between VE and IQs or neuropsychological deficits reduces the possibility that the processes responsible for these deficits are the main determinants of dilatation. Such processes, however, might partially contribute to enlargement during the temporal course of schizophrenia, a possibility that may lead to discrepant results if patient samples with different illness durations are considered.

On the other hand, the high correlation between LNNB scores and patients' verbal and performance IQs or their educational level seems to indicate that LNNB performance may depend more on the subject's ability to fully understand the instructions than on the degree of organic CNS impairment. This strengthens the need for every attempt to correlate neuropsychological battery results with other variables to take simultaneous account of the effects of intellectual and educational factors.

In our investigation, the only D tactile scale remained correlated with VBR after such control procedures were undertaken. This suggests that the impairment of a specific somatosensory neuropsychological function may be a biologically relevant finding in schizophrenics with VE.

The picture that emerges from our analysis is that the impairment revealed by LNNB does not represent an exclusive index of organic brain damage but, rather, indicates a more global functional impairment to which both organic (as expressed by CT scan abnormalities) and other variables (that may or may not correlate with structural changes in the brain) contribute.

Clinical outcome of neuroleptic treatment

Several studies have addressed the issue of whether VE has a prognostic value in schizophrenia and special attention has been given to the relationship between cerebral morphological abnormalities and response to neuroleptic treatment.

A number of prospective studies (32,46,47,48) have reported that during neuroleptic therapy patients with normal ventricular size show a more consistent pattern of improvement than those with enlargement, a finding that may imply that the classical dopaminergic hypothesis may not be applicable to schizophrenia accompanied by evidence or brain atrophy (12,49). But there also have been some negative results (38,39), especially in retrospective studies. Moreover, all these studies were based on brief clinical observations, while information about the long-term outcome of schizophrenic patients with and without VE were lacking.

Considering the relevance of the problem both for clinical and research purposes, we undertook a long-term follow-up study of 25 schizophrenic patients (15 men, 10 women; mean age 32.1 $\pm$ 9.9 years) who had been hospitalized for an acute psychotic exacerbation while drug-free.

Nine patients had VE and 16 had normal VBR values. At admission, BPRS scores between the two groups did not significantly differ, even if they were slightly higher in the "normal ventricles" group (65 $\pm$ 14.9 vs 55.7 $\pm$ 10.7; t=.82, p=NS).

As a group, patients with normal ventricles showed a significantly better clinical outcome, as reflected by a progressive and significant reduction of their BPRS scores over time (X_r^2 Friedmann=17.47, p<.01), while those with enlarged ventricles did not show any improvement (X_r^2 Friedmann=7.38, p=NS) (figure 2).

In patients with normal ventricular size, each monthly BPRS rating was lower than the baseline drug-free value at the .05 to .01 significance level. On the other hand, the only significant difference for patients with enlargement was that between the admission rating and the one obtained a year afterwards (t=2.2, p<.05) (fig.2).

Further, the clinical outcome, as reflected by the percent change of the baseline BPRS scores, was significantly better in patients with normal sized ventricles at each monthly observation.

Besides confirming previous results on the low efficacy of "acute" neuroleptic treatment in patients with VE (46,47,48), these findings also suggest a notably different long-term prognosis for neuroleptic-treated patients with and without ventricular dilatation.

Obviously, before drawing any firm conclusion from our findings, there is need to rigorously control for those non-pharmacological factors that are known to affect the long-term outcome of schizophrenia and may have been differently represented in patients with and without enlargement. If this analysis confirms these preliminary results, we will then have direct proof of the hypothesis that the CT finding of VE may distinguish a form of schizophrenia with a more severe clinical course and negative outcome.

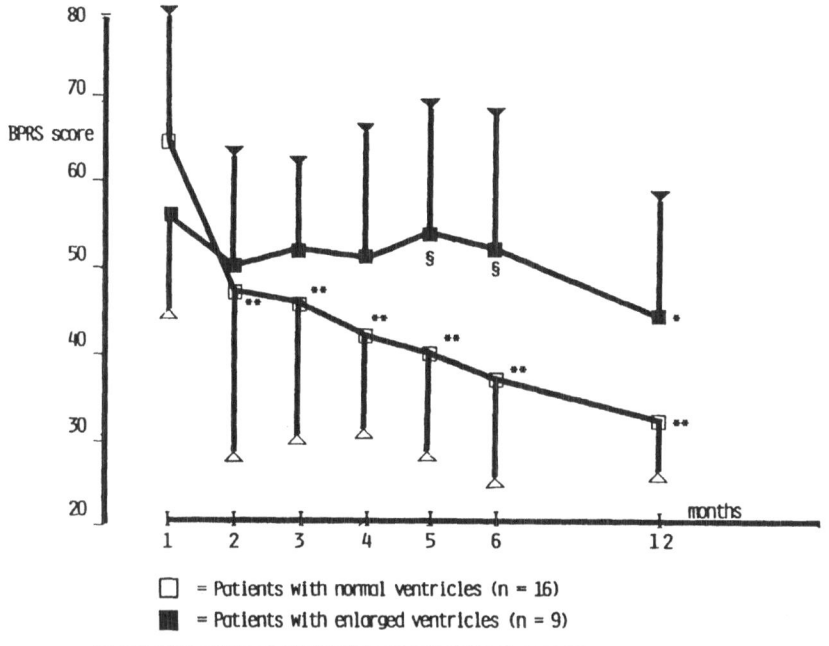

FIGURE 2   Twelve-month-follow-up study of neuroleptic treated schizophrenics with and without ventricular enlargement. A progressive and significant clinical improvement was found in patients with normal ventricles (X^2_r Friedmann = 17.47, p<.01), while patients with VE did not show such a favourable outcome (X^2_r Friedmann = 7.38, p=NS)

81

Further, our patients' poor response to neuroleptic drugs reinforces the idea that the dopamine receptor blockade is not in itself a factor capable of provoking symptom remission in patients with this CT abnormality. This is in line with a number of biochemical and pharmacological studies suggesting that the increase of central dopaminergic activity, the classical dopamine hypothesis of schizophrenia, is not a reasonable feature of schizophrenia with VE (for a review see 11, 12). Both alternative causative hypotheses and therapeutic interventions seem warranted for this form of illness.

Family history of schizophrenia

The question of the relationship between brain morphological abnormalities and the genetic predisposition to schizophrenia is still controversial.

Reveley et al. (50) were able to demonstrate that cerebral ventricular size is under a high degree of genetic control among normal twins, but among monozygotic pairs discordant for schizophrenia they found a greater intrapair difference, with the schizophrenic twins having consistently larger cerebral ventricles than both their normal co-twins and normal controls. This suggested the possibility that VE reflects the presence of an environmental process. Pointing in the same direction is the Schulsinger and coworkers' observation of larger lateral and third ventricles in schizophrenic than in "borderline" or normal chidren of schizophrenic mothers (26).

These findings might lead one to expect that schizophrenics with higher genetic predisposition to the disease (those with other ill family members) would be less likely to have cerebromorphological abnormalities.

Whether "familial" and "sporadic" schizophrenia actually do differ relative to the incidence of brain atrophy, however, is still open question. Examining schizophrenic patients from monozygotic twin pairs Reveley et al. (51) found ventricular volume to be significantly higher in those patients without a known genetic predisposition for the disease. Analogous conclusions were reached by Oxiensterna et al. (53) and Turner et al. (54). On the other hand, neither Weinberger et al. (54) nor Tanaka et al. (55) were able to find any association between ventricular size and FH of schizophrenia. Nasrallah et al. (39) even found increased VBR in patients with a positive FH of the disease. An intermediate and interesting result was reported by Owens et al. (40), who found a curvilinear relationship between VBR and FH, the more "normal" the VBR the higher the incidence of schizophrenic cases in the patients' families, while patients with both abnormally large or very small ventricles had a greatly reduced familial load for schizophrenia.

It is difficult to explain such discrepant results, but size and characteristics of samples on the one hand and the criteria for defining positive vs negative family history may have had a major role.

We reanalyzed this problem in 153 patients, 21 of whom had at least one first degree relative suffering from schizophrenia (FH+), and 132 had no cases of schizophrenia among their 1st and 2nd degree relatives (FH-).

The mean VBR of FH+ patients (4.2 ± 2) was significantly lower than that of FH- ones (5.7 ± 3.2; t=2.0, p<.05).

Furthermore, 27% (35/132) of the FH- patients had enlarged ventricles vs only 14% (3/21) of the FH+ ones. This difference did not reach, however, statistical significance (X^2 = 1.5, p=NS).

Finding larger ventricles in patients without a family history of schizophrenia clearly suggests that, in spite of the fact that VBR is under genetic control (50), VE is not likely to represent the inherited abnormality predisposing subjects to becoming schizophrenic and supports the hypothesis that it may mainly have environmental determinants.

On the other hand, the observation that schizophrenic patients with and without a family history of the disease differ in a such relevant feature as the presence of brain structural abnormalities does support the validity of the familial/sporadic subdivision of schizophrenia as a heuristic device for separating biologically and even etiologically different subtypes of illness (56).

In discussing their results, Reveley et al. (50,51) hypothesized that when the genetic predisposition is high, subjects may develop schizophrenia without any precipitating factor, and that when it is lower the apperance of the disease may require an environmental factor determining brain structural abnormalities that may be detected in CT scans as central or cortical atrophies. We still do not know whether VE is actually relevant to the pathogenesis of schizophrenia, but our findings would support such a hypothesis.

The heuristic relevance of the problem of the relationship between putative brain damage in schizophrenic patients and a genetic predisposition to the disease suggests the need for further studies in this area. Studies assessing the morbid risk for schizophrenia among families of schizophrenic patients with and without cerebral VE seem particularly warranted.

Immunogenetic profile
Since the first study in 1974 (57), a number of associations between schizophrenia and specific HLA antigens have been reported. These associations, however, appear to be weak and inconsistent in the various study populations, a fact attributed to the heterogeneity of schizophrenic disorders. In order to reduce such heterogeneity, Luchins et al. (58) subgrouped their schizophrenic sample according to the presence of brain atrophy, as determined by CT demonstration of cortical or cerebellar vermal atrophy or lateral VE. They found a significant increase of the HLA-A2 antigen in black schizophrenics with no evidence of atrophy, and a similar trend among white patients and concluded that "dividing schizophrenia along these lines" may help to reduce the biological heterogeneity of the disorder.

In a subsequent study, Frangos et al. (59) were unable to replicate this finding in a caucasian schizophrenic sample.

Our interest in better understanding the biological characteristics of patients with and without brain abnormalities led us to reevaluate the HLA antigen distibution among them (6, 14).

Patients with enlarged ventricles (n=11) showed an increased frequency of the HLA-B12 antigen (27.3% vs 4.7%; X^2=7.4, p<.01) while those with normal sized ventricles (n=53) had increased frequencies of HLA-Aw23 (13.2% vs 3.2%; X^2=4.9, p=.03), -Bw51 (26.4% vs 10.7%; X^2=7.4, p<.01) and, though not significantly, of the -A2 antigen (56.6% vs 42.7%; X^2=3.1, p<.1). When compared directly, patients with and without enlargement differed only in their HLA-B12 antigen frequency (27.3% vs 1.9%; X^2=6.2, p<.01). In our sample, therefore, a schizophrenic patient with the HLA-B12 antigen had a risk of showing cerebral VE that was 17,3 times that of one without this antigen.

While our results confirm previous findings regarding a somewhat higher incidence of the A2 antigen among schizophrenics without brain atrophy, we found some other previously unnoted HLA differences between patients with and without VE.

It should be noted that we restricted our definition of brain atrophy to the presence or absence of lateral VE and not to other non-central atrophy signs, since we do not presently know whether the atrophy of different cerebral structures can be assimilated or may result from different pathologic processes. This may account for the discrepancies between our own and previous results in this area and suggests that assimilating abnormalities of different cerebral structures may obscure significant findings.

We really do not know what the significance of the specific HLA associations may be. Assuming the HLA system as a marker of the genetic transmission of diseases one might speculate that schizophrenia with and without ventricular dilatation may have different genetic-familial characteristics, a hypothesis supported by our own and other available familial data (51,52,53). On the other hand, association with HLA antigens has been claimed to suggest involvement of the immune system in the pathogenesis of a disease, that may be different in the CT schizophrenic subgroups. More parsimoniously, the presence of a specific HLA antigen (B12 in our case) may give a subject a higher vulnerability to brain morphological abnormalities, while other antigens may protect him against such an occurence.

Whatever its meaning, finding different and specific HLA associations in patients with and without VE strongly suggests that the CT abnormality distinguishes biologically distinct and more homogeneous subtypes of schizophrenia.

Smooth Pursuit Eye Movements
Impaired smooth pursuit eye movements (SPEM), assumed to reflect impairment of the higher cortical centers, have been found in schizophrenic patients in a series of studies (22, 60).
Two studies have specifically addressed the question of the relationship between SPEM measures and cerebral morphological characteristics in schizophrenia. Weinberger and Wyatt (49) reported more disordered eye movements in patients with CT scan abnormalities, while Bartfai et al. (61) noted a weak, non-significant trend for patients with the most impaired eye tracking to have the largest (but normal) linear measures of anterior horn width, but an inverse correlation between SPEM ratings and cortical atrophy scores.

As part of a larger study of the SPEM characteristics in schizophrenia spectrum disorders (62), we were also interested in looking at possible correlations between SPEM measures and VBR.

Thirty patients (25 men, 5 women; mean age 25.8 ± 7.3 years) had both CT scan and SPEM ratings available. Twenty-five were schizophrenics and 5 had personality disorders of the schizophrenic spectrum. We decided not to separate the two diagnostic groups both because schizophrenia spectrum personality disorder patients behaved akin to schizophrenics on the eye tracking task and because both groups had comparable figures of VE. VBR values were first considered as a source of variation on SPEM measures (qualitative and quantitative ratings) and then SPEM measures were compared in patients with and without VE.

As a covariate, VBR was inversely and highly correlated with both qualitative and quantitative SPEM scores (F=8.5, df 1, p=008 and F=3.96, df 1, p=.05). Further, eye tracking performance was significantly worse in patients with normal ventricular size (mean ratings: qualitative=3.67, quantitative=5.04) than in those with VE (2.77 and 3.66, respectively).

These results are in striking contrast to those of Weinberger and Wyatt (49), even though they are not directly comparable because of our inclusion of patients with spectrum disorders, and are more compatible with those of Bartfai et al. (61), in that they found some inverse relationship between SPEM and atrophy scores.

Although they need replication, these findings suggest that the mechanism/s underlying ventricular dilatation and poor performance in the eye tracking task are different and that these abnormalities may mark distinct subgroups of patients within the schizophrenia spectrum. In our sample, there was very little overlap between VBR and SPEM abnormalities. Ten subjects had abnormal SPEM measures but normal ventricular size, 10 had enlarged ventricles but normal eye tracking performances and only 2 showed abnormal values both for SPEM and VBR measurements.

In attempting to understand this finding, it must be noted that SPEM dysfunction has been interpreted as representing a trait marker possibly related to a genetic vulnerability to schizophrenia (22,60,63). This hypothesis relies on the finding of eye tracking anomalies in no less than 50% of the relatives of the schizophrenics studied, regardless of whether they had any psychopathology (63). We ourselves observed a significant relationship between poor SPEM performance and the presence of secondary cases of schizophrenia among healthy control subjects.

If the hypothesis that SPEM dysfunction is a biological marker of the individual's genetic susceptibility to schizophrenic illness is definitely proven, the inverse correlation between SPEM and VBR measures would strengthen the possibility that VE marks a form of schizophrenia with predominantly environmental determinants or that the CT anomaly is an environmental process itself. This would add to the growing body of evidence relative to the much higher incidence of VE in "sporadic" cases of schizophrenia.

CONCLUSIONS

The results of the various studies described require also to be considered globally in relation to the main question of whether or not it is possible to identify empirically valid and reliable subtypes of schizophrenic disorders on the basis of patient differences in ventricular size. In spite of the shadows still present, the answer to the query is that, phenotypically, VE aggregates in a particular subtype of schizophrenic patients.

Two main prototypes of schizophrenic patients, those with and without VE have emerged from our data. These groups may be distinguished not only in terms of SPEM, HLA typing, specific symptoms and neuropsychological characteristics, but also on the basis of other variables more relevant to treatment. For example, the fact that schizophrenia is differentially present in the families of patients with and without VE implies that the risk for new cases of schizophrenia in schizophrenics' relatives should be differentially assessed according to the neuromorphological characteristics of the proband. Similarly, the fact that patients with VE scarcely benefited from long-term treatment with conventional antipsychotic drugs makes them obvious first-choice candidates for new experimental alternative treatments.

The multifaceted and distinctive picture of patients with normal and abnormal ventricular sizes also warrants some further considerations.

The first of these is that evidence of a substantial pathogenetic discontinuity between the two patient subgroups contrasts with their marked overlapping at a purely symptomatological level.

The second consideration is that even in the case of the most straightforward associations we are still dealing with partial associations, since there is a considerable degree of overlap of the characteristics of the two neuromorphologically distinct subgroups of schizophrenic patients. Within this framework, reducing the inherent heterogeneity of schizophrenia to a simple dichotomy of patients on the basis of their ventricular size is a useful but certain reductive simplification. On the other hand, a continuum of quantitative differences seems more plausible than "all or nothing" distinctions.

The third consideration is that we do not have sufficient knowledge about the real degree of independence or dependence among the several variables that were associated with VE. These two possibilities have different implications. If a situation of substantial independence were the case, the subgroups identified as having normal and abnormal ventricular size would have a poor inherent validity since each of them could then be further subgrouped on the basis of the variables that were found to be independent. Instead, if a situation of substantial dependence were the case, it would be plausible to maintain that the two neuromorphological typologies acceptably identify two main distinct subtypes of schizophrenia with sufficient accuracy and reliability.

Obviously, definite answers in this regard will be available only after we have esplicitly established whether we are witnessing first or second order associations and whether additive or interactive effects among the variables associated with ventricular size are at play. When this area of uncertainty is reduced the researchers will have much less difficulty and be more precise in identifying homogeneous subgroups of patients, an indinspensable prerequisite for a serious challenge to the problem of the etio-pathogenesis of schizophrenia.

ACKNOWLEDGEMENTS

These researches were supported in part by the C.N.R. grants n. 85.00751.56 and 86.01720.56.
The Authors wish to thank Dr. June Shmelzer La Rosa for her precious assistance in revising the manuscript.

REFERENCES

1.   Stevens, JR (1982). Neuropathology of schizophrenia.Arch. Gen. Psychiatry, 39, 1131
2.   Jacobi, W and Winkler, H (1927).Encephalographische studien an chronisch schizophrenen.   Archiv   fur   Psychiatrie   und Nervenkrankenheit, 81, 299
3.   Huber,   C   (1957).   Pneumoencephalographische   und psychopathologische bilder bei endogen psychosen. (Berlin: Springer Verlag)
4.   Cazzullo, CL (1963). Biological and clinical studies on schizophrenia   related   to   pharmacological   treatment.   Rec. Adv.Biol.Psych. 5, 114
5.   Dolan, RJ, Calloway, SP and Mann, AH (1985). Cerebral ventricular size in depressed subjects. Psychol. Med, 15, 873
6.   Vita, A , Sacchetti, E, Calzeroni, A, Invernizzi, G and Cazzullo, CL (1986). Computed tomography in psychiatric disorders. In: Reisner, T, Binder, H, Deisenhammer, E, (eds) "Advances in Neuroimaging". p.243 (Vienna: Verlag der Wiener Medizinischen Akadamie)
7.   Datlof, S, Coleman, PD, Forbes, GB and Kreipe, RE (1986). Ventricular dilatation on CAT scans of patients with anorexia nervosa. Am. J. Psychiatry, 143, 96
8.   Damasio, H, Maurer, RG, Damasio, AR and Chui, H (1980). Computerized tomographic scan findings in patients with autistic behavior. Arch. Neurol., 37, 504
9.   Bird, JM (1982) Computerized tomography, atrophy and dementia: a review. Prog. Neurobiol., 19, 91
10   Johnstone, EC, Crow, TJ, Frith, CD, Husband, J and Creel, L (1976). Cerebral ventricular size and cognitive impairment in chronic schizophrenia. Lancet ii, 924
11.   Weinberger, DR, Wagner, RL and Wyatt, RJ (1983). Neuropathological studies of schizophrenia: a selective review. Schizophrenia Bull., 9, 193
12.   Goetz, KL and Van Kammen, DP (1986). Computerized axial tomography scans and subtypes of schizophrenia. J. Nerv. Ment. Dis., 174, 31

13   Cazzullo, CL, Vita, A, Sacchetti, E, Invernizzi, G, Maffei, C, Ciussani, S, Alciati, A, Pennati, A and Conte, G (1985). Cerebral ventricular size in diagnostic subtypes of schizophrenia and in schizophreniform disorder. In: Cazzullo, CL, Invernizzi, G, (eds) "Schizophrenia: An Integrative View". p.274 (London: John Libbey)

14.   Cazzullo, CL, Sacchetti, E, Vita, A, Illeni, MT, Bellodi, L, Maffei, C, Alciati, A, Bertrando, P, Calzeroni, A, Ciussani, S, Conte, G, Pennati, A and Invernizzi, G (1986). Cerebral ventricular size in schizophrenic spectrum disorders: relationship with clinical, neuropsychological and immunogenetic variables. In: Shagass, C, et al. (eds) "Biological Psychiatry 1985". p.1060. (Amsterdam: Elsevier Science Publ.)

15.   American Psychiatric Association (1980). Diagnostic and Statistical Manual of Mental Disorders. 3rd Edition, (Washington DC: A.P.A.)

16.   Benes, F, Sunderland, P, Jones, BD, LeMay, M, Cohen, BM and Lipinski, JF (1982). Normal ventricles in young schizophrenics. Br. J. Psychiatry, 141, 90

17.   Synek, V and Reuben, JR (1976). The ventricular brain ratio using planimetric measurement of EMI scans. Br. J. Radiology, 49, 233

18.   Penn, RD, Belanger, MG and Yasnoff, WA (1978). Ventricular volume in man computed from CAT scans. Ann. Neurol. 3, 216

19.   Golden, CJ, Hammeke, TA and Purisch, DA (1980). The Luria Nebraska Neuropsychological Battery. Manual. (Los Angeles: Western Psycological Service)

20.   Overall, JE and Gorham, DR (1962).The brief psychiatric rating scale. Psychol Rep., 10, 799

21.   Terasaki, P and McClelland, JD (1964). Microdoplet assay of human serum cytotoxins. Nature. 204, 998

22.   Shagass, C, Roemer, RA and Amedeo, M (1974). Eye tracking performance in psychiatric patients. Biol. Psychiatry, 9, 254

23.   Weinberger, DR, DeLisi, LE, Perman, G, Targum, S and Wyatt, RJ (1982). CT scans in schizophreniform disorder and other acute psychiatric patients. Arch. Gen. Psychiatry, 39, 778

24.   Nyback, H, Berggren, BM and Hindmarsh, T (1982). Computed tomography of the brain in patients with acute psychosis and in healthy volunteers. Acta Psychiat. Scand., 65, 29

25.   Vita, A, Sacchetti, E, Calzeroni, A, Battaglia, M, Bellodi, L, Invernizzi, G and Cazzullo, Cl (1986). Cerebral ventricular enlargement in personality disorders. Presented at the Regional Symposium of the W.P.A.. "Psychiatry and its related disciplines", Copenhagen, 19-22 August 1986. Astract Book F395, p.198

26.   Schulsinger, F, Parnas, J, Petersen, ET, Teasdale, TW, Mednick, SA, Moller, L and Silverton, L (1984). Cerebral ventricular size in the offspring of schizophrenic mothers. Arch. Gen. Psychiatry, 41, 602

27.   Owen, MJ, Lewis, SW and Murray, RM (this volume). Obstetric complications and cerebral abnormalities in schizophrenia

28.   Sacchetti, E, Vita, A, Battaglia, M, Calzeroni, A, Conte, G, Invernizzi, G and Cazzullo, CL (this volume). Season of birth and cerebral ventricular enlargement in schizophrenia

29.   Woods, BT and Wolf, J (1983). A reconsideration of the

relation of ventricular enlargement to duration of illness in schizophrenia. Am. J. Psychiatry, 140, 1564

30. Andreasen, NC, Smith, MR, Jacobs, CG, Dennert, JW and Scott,OA (1982). Ventricular enlargement in schizophrenia: definition and prevalence. Am. J. Psychiatry, 139, 292

31. Kraepelin, E (1919). Dementia Praecox and Paraphrenia. Edited and translated by R.N. Barclay (Edinburgh: E&S Livingstone)

32. Jeste, DV, Kleinman, JE, Potkin, SG (1982). Ex uno multi: subtyping the schizophrenic syndrome. Biol. Psychiatry, 17, 199

33. Boronow, J, Pickar, D, Ninan, PT, Roy, A, Hommer, D, Linnoila, M and Paul, SM (1985). Atrophy limited to the third ventricle in chronic schizophrenia. Arch. Gen. Psychiatry, 42, 266

34. Frangos, EG and Athanassenas, G (1982). Differences in lateral brain ventricular size among various types of schizophrenics. Acta Psychiat. Scand, 66, 459

35. Nasrallah, HA, Jacoby, CG, McCalley-Whitters, M and Kuperman, S (1982). Cerebral ventricular enlargement in subtypes of chronic schizophrenia. Arch. Gen. Psychiatry, 39, 774

36. Kling, AS, Kurtz, N, Tachiki, K and Orzeck, A (1982/83). CT scans in sub-groups of chronic schizophrenics. J. Psychiat. Res., 7, 375

37. Andreasen, NC, Olsen, SA, Dennert, JW and Smith, MR (1982). Ventricular enlargement in schizophrenia: relationship to positive and negative symptoms. Am J. Psychiatry, 139, 297

38. Williams, AO, Reveley, MA, Kolakowska, T, Ardern, M and Mandelbrote, BM (1985). Schizophrenia with good and poor outcome. II: Cerebral ventricular size and its clinical significance. Br. J. Psychiatry, 145, 239

39. Nasrallah, HA, Kuperman, S, Hamra, BJ and McCalley-Whitters, M (1983). Clinical differences between schizophrenic patients with and without large cerebral ventricles. J. Clin. Psychiatry, 44, 407

40. Owens, DGC, Johnstone, EL, Crow, TJ, Frith, CD. Jagoe, JR and Kreel, L (1985). Lateral ventricular size in schizophrenia: relationship to the disease process and its clinical manifestations. Psychol. Med., 15, 27

41. Bishop, JR, Golden, CJ and MacInnes, WD (1983). The BPRS in assessing symptom correlates of cerebral ventricular enlargement in acute and chronic schizophrenia. Psychiatry Res., 9, 225

42. Crow, TJ (1980). Molecular pathology of schizophrenia. More than one disease process? Br. Med.J., 280, 66

43. Chelune, JC (1982).A reexamination of the relationship between Luria-Nebraska and Halstead-Reitan batteries: overlap with the WAIS. J. Consult. Clin. Psychol., 50, 578

44. Maffei, C, Luoni, P, Vita, A, Bertrando, P, Cesana, B, Ciussani, S and Sacchetti, E (1987). The Luria Nebraska neuropsychological battery in schizophrenic disorders. Relationship to neuromorphological, intellectual and educational variables. New Trends Exp. Clin. Psychiatry, 3, 25

45. Golden, CJ, Moses, JA, Zelazowski, R, Graber, B, Zatz, LM, Horvath, TB and Berger, PA (1980). Cerebral ventricular size and neuropsychological impairment in young chronic schizophrenics. Arch. Gen. Psychiatry, 37, 619

46. Weinberger, DR, Bigelow, LB, Kleinman, JE, Klein, ST,

Rosenblatt, JE and Wyatt, RJ (1980). Cerebral ventricular enlargement in chronic schizophrenia: an association with poor response to treatment. Arch. Gen. Psychiatry, 37, 11

47.   Schulz, SC, Sinicrope, P, Kishore, P and Friedel, RO (1983). Treatment response and ventricular brain enlargement in young schizophrenic patients. Psychopharmacol. Bull, 19, 510

48.   Luchins, DJ, Lewine, RJ and Meltzer, HY (1984). Lateral ventricular size, psychopathology and medication response in the psychoses. Biol. Psychiatry, 19, 29

49.   Weinberger, DR and Wyatt, RJ (1982). Cerebral ventricular size: a biological marker for subtyping chronic schizophrenia. In:Usdin, E, Hanin, I, (eds) "Biological Markers in Psychiatry and Neurology". p.502 (New York: Pergamon Press)

50.   Reveley, AM, Reveley, MA, Clifford, CA and Murray, RM (1982). Cerebral ventricular size in twins discordant for schizophrenia. Lancet i, 540

51.   Reveley, AM, Reveley, MA and Murray, RM (1984). Cerebral ventricular enlargement in non-genetic schizophrenia: a controlled study. Br. J. Psychiatry, 144, 89

52.   Oxenstierna, G, Bergstrand, G, Bjerkenstedt, L, Sedvall, G and Wik, G (1984). Evidence of disturbed CSF circulation and brain atrophy in cases of schizophrenic psychosis. Br. J. Psychiatry, 144, 645

53.   Turner, SW, Toone, BK and Brett-Jones, JR (1986). Computed tomographic scan changes in early schizophrenia. Preliminary findings. Psychol. Med., 16, 219

54.   Weinberger, DR, DeLisi, LE, Neophytides, AN and Wyatt, RJ (1981). Familial aspects of CT scan abnormalities in chronic schizophrenic patients. Psychiatry Res., 4, 65

55.   Tanaka, Y, Hazama, H, Kawahara, R and Kobayashi, K (1981). Computerized tomography of the brain in schizophrenic patients. Acta Psychiat. Scand., 63, 191

56.   Murray, RM, Lewis, SW and Reveley, AM (1985). Towards an aetiological classification of schizophrenia. Lancet i, 1023

57.   Cazzullo, CL, Smeraldi, E and Penati, G (1974). The leukocyte antigenic system HLA as a possible genetic marker of schizophrenia. Br. J. Psychiatry, 125, 25

58.   Luchins, D, Torrey, EF, Weinberger, DR, Zalcman, S, DeLisi, L, Johnson, A, Rogentine, N and Wyatt, RJ (1980). HLA antigens in schizophrenia. Differences between patients with and without evidence of brain atrophy. Br. J. Psychiatry, 136, 243

59.   Frangos,E, Renieri-Livieratou, N and Athanassenas, G (1980). HLA antigens in schizophrenia: No difference between patients with and without evidence of brain atrophy. Br. J. Psychiatry, 140, 607

60.   Holzman, PS and Levy, DL (1977). Smooth pursuit eye movements and functional psychoses: a review. Schizophrenia Bull., 3, 15

61.   Bartfai, A, Lewander, SE, Nyback, H, Berggren, BM and Schalling, D (1985). Smooth pursuit eye tracking, neuropsychological test performance, and computed tomography in schizophrenia. Psychiatry Res., 15, 49

62.   Smeraldi, E, Gambini, O, Bellodi, L, Sacchetti, E, Vita, A, Di Rosa, M and Cazzullo, CL (in press). Combined measures of smooth pursuit eye movements and ventricular brain ratio in subjects with

schizophrenic disorders. Psychiatry Res.

63.   Holzman, PS, Proctor, LR, Levy, DL, Yasillo, NJ, Heltzer, NY and Hurt, SW (1974). Eye tracking dysfunction in schizophrenic patients and their relatives. Arch. Gen. Psychiatry, 31, 143

9
Season of birth and cerebral ventricular enlargement in schizophrenia

E. Sacchetti, A. Vita, M. Battaglia,
A. Calzeroni, G. Conte, G. Invernizzi
and C.L. Cazzullo

INTRODUCTION

The theme of the seasonal tendency of schizophrenic births certainly
is not new for experimental psychiatry. The tendency was first noted
about 60 years ago by Tramer (1), who noted an excess of
schizophrenic births during the winter and early spring. Following
this, a large number of studies were undertaken, taking thousands of
cases into consideration (for references, see 2,3,4). Undoubtedly,
some uncertainty still exists about the role each month plays in
generating this asymmetrical distribution of schizophrenic births as
well as about the exact dimensions of this phenomenon. Nonetheless,
the association between the future condition of schizophrenia and
birth in the period of the year between December and April (in the
northern hemisphere) is no longer the subject of doubt.

Two main hypotheses tentatively have been proposed to explain
this finding (see 2,3).

The first of these, still the most accredited, holds that a
seasonal increased risk of exposure to factors such as viral
infections, obstetrical and perinatal complications and extreme
temperatures facilitates the structuralization of organic cerebral
damage that results in a progressive behavorial disorganization
until the development of a typical schizophrenic disorder.

On the other hand, the second hypothesis stipulates that an
increased propensity of parents to conceive during the spring and
early summer is at the base of the monthly asymetrical fluctuation
of schizophrenic births.

In the case of both hypotheses, however, the arguments both pro
and con are still highly speculative. From this situation comes the
impetus to promote studies specifically addressed to giving more
support to one or the other hypothesis.

A relevant challenge to hypotheses based on the presumption of
seasonally varying risk for brain damage could be to demonstrate
that birth in a determinate period of the year increases the
probability that schizophrenic adults have a demonstrable
neuromorphological anomaly such as ventricular enlargement as
assessed by CT scans.

We already know that ventricular enlargement occurs in

schizophrenic patients at a mean frequency of about one-third to one-quarter of all schizophrenic cases (5,6). This appears to be compatible with the 5-20% increased risk of schizophrenia generally reported for individuals born in the winter and early spring.

Further, the timing of the appearance of ventricular enlargement seems to be localized backwards in time, at a point antecedent to the onset of schizophrenic disorder and, perhaps, during an earlier stage of brain development. Although there is no uniformity of opinion, this neuromorphologic abnormality sometimes has been related to premorbid functioning (7), to low birth weight (8), hours of labour (9) and obstetrical complications (10,11). Further, ventricular enlargement is detectable in early schizophrenic patients, in patients with schizophreniform disorders and in individuals with personality disorders of the schizophrenic spectrum, a condition predisposing the individual to the subsequent development of schizophrenia (5,10,12,13,14,15).

In light of the above and the fact that this experimental approach has been completely neglected, we decided to examine one of our very preliminary observations (16) and to study whether or not schizophrenic patients born between December and April present an excessive number of cases of ventricular enlargement.

SUBJECTS AND METHODS

One hundred-fifty-five patients (mean age 26.2 ± 7.2 years) who had undergone CT scan examination entered the study. Inclusion criteria were: 1) a diagnosis of schizophrenia according to the DSM-III criteria (17), 2) no evidence of neurological or medical illnesses, 3) no history of drug abuse or heavy alcohol intake, head trauma with loss of consciousness and seizures; 4) no intake of steroids in the 3 months before CT scan.

The CT control group consisted of 57 healthy subjects who had had CT scan to exclude brain damage from minor accidental head trauma without loss of consciousness. Their characteristics are reported elsewhere (5).

Cerebral ventricular size was measured on the CT slice showing the lateral ventricles at their largest by use of a manual planimetric grid method (for description see 5), and expressed as ventricular brain ratio (VBR) (18). Ventricular enlargement (VE) was expressed as a VBR exceeding the mean VBR for controls by more than 2SD.

RESULTS AND COMMENTS

In the present sample, 115 patients, the majority, had VBR values overlapping those found in controls and 40, i.e. 25.8% of the entire population, had ventricular enlargement (VBR > 7.5).

The cases with and without ventricular enlargement were then divided into those born between December and April and those born in the other months of the year. This division was undertaken regardless of whether the patients were subtyped as disorganized (n=83), paranoid (n=28) or undifferentiated (n=44) or whether they were chronic (n=120) or subchronic (n=35), since all these variables

94

were equally distributed among the two neuromorphological subgroups.

Contrarily, there was no random distribution of an individual's chances for ventricular enlargement if he was born during the two different periods of the year. As many as 25 of the 69 patients born in the winter and early spring (i.e. 36%), showed pathological VBR values, while only 15 of the patients born between May and November, that is 16% out of the total of 86, had the same abnormality ($X^2 =$ 7.05; p = .008).

The finding that patients born between December and April have a ventricular enlargement risk 2.7 times greater than those born during the rest of the year extends to a sizeable number the patients we studied preliminarily (16) and, to the best of our knowledge, provides the first adequate experimental basis indicating a link between brain damage and seasonality of schizophrenic births.

This result was not entirely unexpected. At least two other indirect factors prepared us for what we found.

The first of these is that one of the possible but not specific factors associated with schizophrenia is birth complications. We know that birth complications are more likely to occur among individuals born in the winter (19,20) and those who have a higher risk of ventricular enlargement (10,11,21).

The second indirect factor relates to familial predisposition to schizophrenia. On the one hand, it seems ever more apparent that ventricular enlargement identifies patients suffering from a more sporadic subform of schizophrenia (5,6,21). On the other hand, there is some evidence that a seasonal birth pattern is more pronounced among patients with a negative family history of schizophrenia (22,23).

Even if data on obstretical complications and familial predispositions to schizophrenia provide a supplementary bridge between ventricular enlargement and seasonality of birth, conflicting evidence certainly exists.

For example, while schizophrenic patients born in the winter seem to suffer from a less chronic disease with a better prognosis than patients born at other times of the year (24,25) there is accumulating evidence suggesting that patients with ventricular enlargement would suffer more impairment with their disorder and not respond to neuroleptics as patients with normal ventricles do (5,6).

Data concerning a possible better prognosis for winter-born schizophrenics, however, are far from conclusive. This possibility has not been validly confirmed in all instances in which it has been investigated (25). Further, if we accept being unmarried as an index of greater and more severe social impairment, demonstrating that "single" patients have a more marked birth seasonality (4) can be used as proof to the contrary.

Anyway, these apparent discrepancies would not represent sufficient proof against the direct demonstration that ventricular enlargement mainly affects that specific subtype of schizophrenics born in the winter or early spring. In fact, by definition one must await "non seasonal" schizophrenic births between December and April at frequencies equal to those of the general population during the same period of the year. Similarly, as far as the causes of ventricular enlargement are concerned, not all of them must necessarily have an equally well-pronounced seasonal pattern, and, at any rate, they cannot have an all or nothing type of distribution in the various periods of the year. Consequently, an incomplete relationship between characteristics associated with ventricular enlargement and those associated with birth between December and early spring is inevitable at this point.

Obviously, the demonstration that patients born in winter and early spring are at higher risk for ventricular enlargement does not represent a point of arrival in our efforts to validate the hypothesis that seasonal factors surrounding birth predispose the individual to a later development of schizophrenia via brain damage. In fact, several imperative questions are awaiting specific responses.

Some of these questions, for example those regarding the possibility that the neuropathological processes triggered by seasonal factors have an immediate rather than a long-term effect on ventricular size or that regarding the complex chain of events that solder brain damage and the appearance of a schizophrenic disorder, undoubtedly are destined to remain a merely heuristic framework because they are beyond our present conceptual and research possibilities.

On the contrary, two other relevant questions seem to have possibilities for investigation in the near future.

The first of these relates to the specific causal role that obstetrical and perinatal complications, viral infections, temperature, nutritional deficiences or other seasonal factors may have on increasing the incidence of cases with ventricular enlargement among schizophrenic patients born in the winter and early spring.

The second question focuses on the need to unequivocally demonstrate whether the link between seasonality of birth and ventricular enlargement is or is not disease-specific for schizophrenia . There already is some evidence that patients with major affective disorders also are susceptible to an analogous trend of birth seasonality (see 2). There also is substantial evidence that a significant proportion of patients with major affective disorders also have ventricular enlargement (26,27).

Given the obvious relevance that definitive answers to these two questions would have for a real improvement of the data on the schizophrenic birth seasonality hypothesis, we would then have a rapid increase of research focused on these specific areas, research not only needed but also predictably rewarding.

ACKNOWLEDGEMENTS

These researches were supported in part by the C.N.R. grants n. 85.00751.56 and 86.01720.56.
The Authors wish to thank Dr. June Shmelzer La Rosa for her precious assistance in revising the manuscript.

REFERENCES

1.   Tramer,   M (1929).   Uber die biologische   Bedeutung des
Geburtsmonates, insbesondere fur die psychose Erkrankung. Schweiz.
Archiv. Neurol. Psychiatr., 24, 17.
2.   Boyd, JH, Pulver, AE and Steward, W (1982).Season of birth:
schizophrenia and bipolar disorder. Schizophrenia Bull., 12, 173.
3.   Torrey,   EF (this volume).   Hypotheses on the seasonality of
schizophrenic births.
4.   Watson, CG, Kucala, T, Tilleskjor, C and Jacobs, L (1984).
Schizophrenic seasonality in relation to the incidence of infectious
diseases and temperature extremes. Arch. Gen. Psychiatry, 41, 85.
5.   Sacchetti, E, Vita, A, Calzeroni, A, Invernizzi, G and
Cazzullo,   CL (this volume).   Neuromorphological   correlates   of
schizophrenic disorders: focus on cerebral ventricular enlargement.
6.   Goetz, KL and Van Kammen, DP (1986). Computerized axial
tomography scans and subtypes of schizophrenia. J. Nerv.   Ment.
Dis., 174, 31.
7.   Weinberger, DR,   Cannon-Spoor, ME, Potkin, SG and Wyatt, RJ
(1980). Poor premorbid adjustment and CT scan abnormalities in
chronic schizophrenia . Am. J. Psychiatry, 137, 1410
8.   Schulsinger, F, Parnas,J, Petersen, ET, Teasdale, TW, Mednick,
SA, Moller, L and Silverton, L (1984). Cerebral ventricular size in
the offspring of schizophrenic mothers: a preliminary study. Arch.
Gen. Psychiatry, 41, 602.
9.   Takahashi, R, Inaba, Y, Inanaga, K, Kato, N, Kumashiro, H,
Nishimura, T, Okuma, T, Otsuki, S, Sakai, T and Sato, T (1981). CT
scanning and the investigation of schizophrenia. In: Perris, C,
Struwe, G and Jansson, B (eds) "Biological Psychiatry 1981". p.259.
(New York: Elsevier/North Holland Biomedical Press).
10.   Turner, SW, Toone, BK and Brett-Jones, JR (1986). Computed
tomographic scan changes in early schizophrenia - preliminary
findings. Psychol. Med., 16, 219.
11.   Owen, MJ, Lewis, SW and Murray, RM (this volume). Obstetric
complications and cerebral abnormalities in schizophrenia.
12.   Schulz, SC, Koller, MM, Kishore, PR, Hamer, RM, Gehl JJ, and
Friedel, RO   (1983). Ventricular enlargement in teenage patients
with schizophrenia spectrum disorder. Am. J. Psychiatry, 140, 1592.
13.   Weinberger, DR, DeLisi, LE, Perman, GP, Targum, S and Wyatt,RJ
(1982). Computed tomography in schizophreniform disorder and other
psychiatric disorders. Arch. Gen. Psychiatry, 39, 778.
14.   Vita, A, Sacchetti, E, Calzeroni, A, Invernizzi, G and
Cazzullo, CL (1986). Computed tomography in psychiatric disorders.
In: Reisner, T, Binder, H and Deisenhammer, E (eds) "Advances in
Neuroimaging", p.243.   (Wien: Verlag der Wiener Medizinischen
Akademie).
15.   Cazzullo, CL, Sacchetti, E, Vita, A, Illeni, MT, Bellodi, L,
Maffei, C, Alciati, A, Bertrando, P, Calzeroni, A, Ciussani, S,
Conte, G, Pennati, A and Invernizzi, G (1985). Cerebral ventricular
size in schizophrenic spectrum disorders: relationship to clinical,
neuropsychological and immunogenetic variables. In: Shagass, C et
al. (eds) "Biological Pschiatry 1985". p.1060. (Amsterdam: Elsevier
Science Publ.)

16.   Sacchetti, E, Vita, A, Ciussani, S, Alciati, A, Conte, G,
Pennati, A, Invernizzi, G and Cazzullo, CL (1984). Season of birth
and cerebral ventricular enlargement in schizophrenia. Satellyte
Meeting of the XIV CINP Congress "Schizophrenia: an integrative
view". Milan, June 15-18, 1984, Book of Abstracts, p.96.
17.   American Psychiatric Association (1980). DSM III:Diagnostic
and Statistical Manual of Mental Disorders, 3rd edition (A.P.A.:
Washington D.C.).
18.   Synek, V and Reuben, JR (1976). The ventricular brain ratio
using planimetric measurements of EMI scans. Br. J. Radiol., 49,
233.
19.   Pasamanick, B, and Jnobloch, H (1985). Seasonal variation in
complications of pregnancy. Obstet. Gynecol., 12, 110.
20.   Perlstein, MA and Hood, PN (1967). Seasonal variation in
congenital cerebral palsy. Devel. Med. Child. Neurol., 9, 673.
21.   Reveley, AM, Reveley, MA and Murray, RM (1984). Cerebral
ventricular enlargement in non-genetic schizophrenia: a controlled
twin study. Brit. J. Pychiatry, 144, 89.
22.   Kinney, DK and Jacobson, B (1978). Environmental factors in
schizophrenia: new adoption study evidence. In :Wynne, LC, Cromwell,
RL and Matthysse, S (eds) "The Nature of Schizophrenia: New
Approaches to Resarch and Treatment". p.38. (New York: J. Wiley).
23.   Shur, E (1982). Season of birth in high and low risk
schizophrenics. Br. J. Psychiatry, 140, 410.
24.   Pulver, AE, Stewart, W, Carpenter, WT jr. and Childs, B (1983).
Risk factors in schizophrenia: season of birth in Maryland, USA.
Br. J. Psychiatry, 143, 389.
25.   Dalen, P (1975). Season of birth: a study of schizophrenia and
other mental disorders (Amsterdam: Elsevier/North Holland Biomedical
Press)
26.   Pearlson, GD and Veroff, AE (1981). Computerized tomographic
scan changes in manic-depressive illness. Lancet ii, 470.
27.   Sacchetti, E, Vita, A, Calzeroni, A, Invernizzi, G and
Cazzullo, CL (in press). Cerebral ventricular enlargement: a marker
for subtyping major affective disorders? Proceedings of the
International Congress "Diagnosis and Treatment of Depression".
Montpellier, May 11-12, 1987.

10
Twin and family studies of ventricular enlargement in schizophrenia

M.A. Reveley, A.M. Reveley and B. Chitkara

Most, but not all, computerised tomographic (CT) scan studies have found ventricular enlargement (VE) in schizophrenia.[1] Overall, perhaps 25% of schizophrenics have some degree of enlargement. Whilst many researchers have attempted to correlate various clinical variables with scan abnormalities, there have been relatively few twin or family studies, which are particularly suited to unravel the relative contributions of genetics and environment in the causation of VE.

TWIN STUDIES

As heredity contributes considerably to a liability to schizophrenia, CT scan abnormalities could reflect an inherited abnormality. However, obstetric complications, infectious agents, head injury or other factors which cause or are secondary to schizophrenia, (for example neuroleptic medication or ECT) could also be responsible for structural changes. In order to partial out these two possible contributing factors, Reveley et al [2], examined the CT scans of 11 pairs of healthy monozygotic (MZ) and 8 pairs of dizygotic (DZ) twins and compared them with a group of 7 pairs of MZ twins discordant for schizophrenia. We hypothesised that if some individuals were genetically predisposed to large ventricles and these were in turn a contributing, but not direct cause of schizophrenia, then both schizophrenic twin and healthy cotwin should have a similarly increased ventricular size. However, if VE were the result of external agents or injury to the affected twin, or were a consequence of the illness itself, then only the schizophrenic twin should show the abnormality.

The control twins were selected from the Institute of Psychiatry normal volunteer twin register and the schizophrenic twins, from the Maudsley Hospital twin register. We used the Maudsley Hospital EMI CT 1010 scanner, which takes cuts 1cm in thickness parallel to the orbitomeatal line, producing a 150 x 160 pixel matrix. The diagnosis of schizophrenia was established by (a) a SADS-L interview,[3] (b) review of the records of all hospital admissions and (c) consultation with relatives. All of the schizophrenics met Research Diagnostic [4] and DSM III [5] criteria for schizophrenia. The scan records were stored on floppy discs and viewed on a computerised video display unit (VDU). Ventricular and intracranial area were determined manually by tracing around their circumference using the VDU pointer, and the ventricular/brain ratio was calculated(VBR=ventricular area/brain x 100).

Intraclass correlations (r) between pairs were also calculated, which gave an approximate measure of heritability, since h^2=r/co-efficient of relationship (co-efficient of relationship = 1 for MZ and 0.5 for DZ twins, since MZ's share 100%, and DZ's, on average, 50% of their genes).

Both largest ventricular area and largest VBR were very highly correlated in the healthy MZ twins (r=0.98, p<0.001, h^2=98%), and the DZ correlation was approximately half the MZ value (R=0.45, ns, h^2=90%). This suggested that ventricular size was under a very high degree of genetic control. We were able to demonstrate the usual high heritability expected for anthropometric measurements, despite possible variability in head position in the scanner and other sources of artefact. However, the correlation for ventricular area and VBR among the discordant pairs were both 0.87, which was significantly (P<0.01) less than that of the normal control twins, because of increased intrapair variability. This implies an additional environmental influence affecting ventricular size in the schizophrenics.

The schizophrenics also had significantly greater ventricular areas and VBR's than their cotwins and the normal controls. There was a non-significant trend for the cotwins to have larger values than normal. Possible explanations are that the schizophrenic process itself or its treatment could have been responsible, as in 6 out of the 7 pairs of MZ twins discordant for schizophrenia, it was the schizophrenic who had the larger ventricles. As the cotwins also had larger ventricles than controls, this suggests that factors consequent upon the illness are unlikely to be the full explanation. Another possibility is that in our series organic damage due to low birth weight and complicated delivery had occurred in 5 of the 14 individuals and included epilepsy, acalculia, and dyslexia. However, such abnormalities were also present in the normal MZ pairs. By choosing twins discordant for schizophrenia we may have assembled a series in which environmental factors were especially important.

In an attempt to further explore the relationship of genetic and environmental influences, we [6] studied a group of 21 schizophrenics of MZ twin birth and 18 controls from 18 healthy MZ pairs. The patients were diagnosed and scanned as in the prior study.[2] In this study we assessed ventricular size by the measurement of total ventricular volume (TVV), using the Maudsley Hospital diagnostic enhancement package software, which enabled us to sum precisely all those pixels within ventricular spaces with CSF density. We were particularly interested in examining the effects of family history and birth complications on ventricular size. Births were classified as "complicated" if they were reported as featuring a breech or difficult forceps delivery, a blue baby, or birth weight less than 3lbs 4ozs (1.48kgs).[7] Prematurity by itself was not considered a birth complication. Family history was regarded as positive if the twin reported and the case notes or relatives confirmed that a first or second degree relative had suffered from a major psychiatric illness.[8] The schizophrenic group as a whole had a greater mean

TVV than the controls, but the difference was not significant after square root transformation of the data (1188 $\pm$ 195 vs 712 $\pm$ 89 voxels, t=1.86, ns). However, the control twins who reported birth complications had very significantly larger TVV than those controls who did not (TVV = 1015 $\pm$ 120 vs 470 $\pm$ 58 voxels, t=4.30, P<0.001). Analysis of variance of $\overline{TVV}$ by sex and reported birth complications using age as a covariant, showed that the effect of birth complications was highly significant (F=16.086, P=0.001). However, the same analysis for the TVV of schizophrenics did not yield a significant result. Many without birth complications showed increased ventricular size. However, all schizophrenics with birth complications had a negative family history, and 5 out of 6 came from pairs discordant for schizophrenia. Analysis of variance of ventricular volume by family history, twin concordance and birth complications(controlling for present age by analysis of covariance) showed family history to be the only significant effect (F=7.48, P=0.016) (family history negative versus family history positive schizophrenics, TVV=1503 $\pm$ 244 vs 558 $\pm$ 151 voxels).

Our hypothesis that enlarged cerebral ventricles would be found only among schizophrenics without obvious genetic predisposition was confirmed: all the schizophrenics with a positive family history had ventricular volumes below the mean of the schizophrenic group as a whole. The association of perinatal trauma and TVV among the controls provides direct evidence of an association between birth complications and increased ventricular size in a normal population. Schulsinger [9] has also noted that among the offspring of schizophrenic mothers, signs of prematurity at birth are associated with increased ventricular size in adult life.

While the proportion of twins reporting birth complications was high (44%) we did not find an association of birth complications and ventricular enlargement among the schizophrenics. Indeed, amongst the 14 schizohrenics with a negative family history, the 6 reporting birth complications had a lower mean TVV (1314 $\pm$ 458 voxels) than the 8 who did not report complicated birth (1644 $\pm$ 581 voxels). Thus, some other, unknown process must have led to ventricular enlargement among the schizophrenics.

A variety of cerebral events may lead to ventricular enlargement, and numerous organic insults or specific genetic syndromes have been associated with schizophrenia. We suggest that a schizophrenic illness may not be the invariable result of any given stress or particular genetic background, but that there is a continuum of liability to schizophrenia, so that those with a ligher "genetic load" may require environmental influence to manifest the illness and even then show less severity, while those with a greater genetic predisposition may become schizophrenic on this basis alone or with a minimal environmental contribution.

In an attempt to further explore what environmental factors may be involved, we [10] examined selected adverse events in 11 pairs of twins discordant for schizophrenia. These included the original 7 pairs,[2] plus an additional 4 pairs. Clinical, CT scanning,and measurement were the same as described previously.[6,11] Adverse environmental events included gunshot wound to the body (1), closed head injury with unconsciousness (2), yaws (1), pyloric stenosis (1), hernia operation as infants (2), and birth anoxia (6). We

101

found that significantly more adverse premorbid events were reported for the schizophrenic members of the 11 twin pairs (t=4.18, P<0.01), suggesting that nonspecific debilitating illness or injury may play a part in precipitating a schizophrenic illness in a predisposed individual. There were no differences in birth weight or any effect of birth order. In only those pairs with a negative family history, the schizophrenics had significantly and consistently (1605 $\pm$ 1040 vs 1191 $\pm$ voxels, t=2.47, P<0.05), larger ventricles than their well, unmedicated cotwins. These results gave further confirmation of the "diathesis-stress" model.

In order to examine the genetic basis of cerebral ventricular volume in a group of normal twins, we examined an extended sample of 18 monozygotic and 18 dizygotic twin pairs and examined heritability values using different methods of calculation. We used the same clinical and diagnostic methods as previously described and used the fixed threshhold method of determining ventricular volume.[11] We used the classical approach to the analysis of twin data and partitioned the total variation in the sample into "between pairs" and "within pairs" variance and a degree of similarity of the MZ and DZ series were compared. The degree of genetic determination, or heritability, can be derived in various ways from twin data but all are subject to bias and depend upon a number of assumptions. For example, genetic and environmental variance are assumed to be the same in both types of twins and no allowance can be made for dominance or gene-environment interaction. The simplest expression for heritability is : (1) $h^2 = r/G$, where r is the intraclass correlation and G is the degree of genetic similarity (assumed to be 0.5 in DZ twins). The expression for heritability suggested by Falconer [12] is (2) $h^2 = 2(r_{mz} - r_{dz})$. The classical twin analysis heritability is derived as (3) $h^2 = I_{adz} - I_{amz}/I_{adz}$, assuming the interpair variance I_e to be comparable. The heritabilities for these different methods were 0.835, 0.852, 0.834 and 0.818 for equations (1) MZ, (1) DZ, (2), and (3), respectively (P=0.031, regardless of method used to calculate heritability). Within this sample we did not attempt to examine the effect of birth complications or perinatal morbidity on heritability, which might have been higher had the calculations been restricted only to twins without birth complications.

FAMILY HISTORY STUDIES OF VENTRICULAR SIZE IN SCHIZOPHRENIA

Other investigators have attempted to replicate in singletons our family history result in twins. Three studies [14-16] with small sample sizes found no relationship, while Nasrallah et al [17] are alone in reporting a higher rate of schizophrenia in the relatives of patients with large compared to those with normal ventricles. Four studies [18-21], in addition to the twin studies [6,10] have found an association of ventricular enlargement with negative family history . Owens et al [22] found a curvilinear relationship between a definite family history of schizophrenia and VBR in 112 patients. They found that a positive family history of schizophrenia was associated with normal ventricular size and that extremes of large and small ventricles were less likely to have a positive family history. This introduces the interesting possibility that extremely small ventricles should be regarded as pathological, as well as

102

extremely large ones.

A particularly unique study by Schulsinger et al [9] examined ventricular size in a prospective longitudinal study of offspring of schizophrenic mothers. Schizophrenics exhibited larger ventricular size, and borderline schizophrenics (DSM III schizotypal), smaller ventricular size than mentally healthy controls. Ventriclar size correlated with premorbidly attained obstetric data. There was no relationship between ventricular size and age, length of psychiatric hospitalisation, drug treatment or ECT. The results suggested support for the diathesis stress model, which considers schizophrenia as a result of deleterious environmental influences acting on a genetic predisposition. Thus, both the schizophrenics and borderline schizophrenics inherited a comparable genetic loading, that was more severe than that of the no mental illness group. Those who remained at the borderline level appeared to have been exceptionally free from potentially detrimental stressors, e.g. perinatal complications, to which the schizophrenics had been exposed. In the absence of a schizophrenic genotype, such detrimental stressors will not produce schizophrenia, but given the necessary genetic predisposition, additional CNS insults increase the likelihood of psychosis, whether schizophrenia or affective disorders, developing.

FAMILY STUDIES OF VENTRICULAR SIZE IN SCHIZOPHRENIA

Weinberger et al [23] examined VBR and measures of cortical structures in patients with schizophrenia, their nonschizophrenic siblings, and healthy siblings from normal sibships. Ventricular size among control siblings was highly correlated ($r=0.73$, $p<0.0015$), if one deviant family were excluded, demonstrating in singletons the genetic component to ventricular size. Within the schizophrenic sibships the patients had significantly larger ventricles than siblings and in every sibship the schizophrenic patient had the largest ventricles even if his were within the normal range. Furthermore, ventricular size in the schizophrenic singletons and their siblings was much more variable than in the control sibships ($F=7.16$, $df=21,16$, $P<0.001$), as we [2] had found in twins. These results suggest that VE is a marker of the clinical illness, and not a trait or genetic marker of vulnerability. Thus VE is the result of environmental, not genetic, factors in schizophrenia.

Ventricular size was significantly greater in the non-schizophrenic siblings (5.5 ± 2.5), as compared to the controls (2.76 ± 1.6, $t=3.65$, $P=0.001$). This result, however, is dissimilar to ours with the schizophrenic cotwins, who while intermediate between controls and schizophrenic twins, were not significantly different from either group. Patients with atrophy had the same prevalence of schizophrenia in their families as patients with normal CT scans, weighing against the link between CT abnormalities and familial or genetic schizophrenia.

In a larger family study, DeLisi et al [24] examined brain lateral ventricular size in 26 schizophrenic subjects from 12 unrelated families, their available well siblings (n=10) and 20 nonpsychotic controls. The mean frontal horn VBR of the schizophrenic subjects was significantly greater than that of both controls, (2.82 vs 2.09, $P=.05$ by 1 tailed t-test), and their nonpsychotic siblings

(2.82 vs 1.83, t=2.2, p=0.03). The VBR as measured through the bodies of the lateral ventricles was also significantly greater in schizophrenics compared to controls (8.25 vs 6.39, P=0.05), but not compared to their non-psychotic siblings.

Of particular interest is the high prevalence of adverse environmental events in this group selected for presumed "genetic" schizophrenia. Four schizophrenic siblings had a history of obstetric complications and 2 of these had the largest frontal horn VBRs. Five schizophrenics and no controls had a history of significant head injury. One had a history of viral meningitis at age 5 years. Analysis of variance showed that for frontal horn VBR, age (P=.004), birth complication (P=.0001), and head injury (P=.0009) were significant sources of variance within the sample. After their variance was accounted for, family (P=0.031) and diagnosis (P=0.052) were still significant factors. For VBR through the lateral ventricles, age (P=.05), and head injuries (P=.003), were significant factors, while family (P=.019) remained significant after the effects of age, birth complications and head injuries were controlled. However, diagnosis, (P=.87), and birth complications (P=.19) were not significant predictors of VBR through the bodies.

The authors suggest that as the increment in ventricular size is observed in families with multiple schizophrenic members, the data suggest that increased ventricular size could represent an inherited vulnerability towards schizophrenia. However, several considerations argue against these conclusions. (1) The most statistically significant effects were environmental. Taken together history of birth complications or head injuries were present in all but one of the 8 schizophrenics with a frontal horn VBR greater than 1 standard deviation beyond the control mean. (2) There was no comparison of familial versus non-familial schizophrenics in this study. (3) If VE represented a genetic diathesis, one would have expected the non-psychotic siblings to have larger VBR than control, which was not the case at either level. The results could equally be interpreted that increased VBR is an illness, rather than a genetic, marker.

The authors also did not find a significant diagnosis effect at the bodies of the lateral ventricles and suggest that technical difficulties with the scanning procedure may have accounted for this. However, Reveley [11] demonstrated that computerised methods, mechanical planimetry and even Evan's ratio could distinguish schizophrenics from controls, regardless of slice, as long as comparable slices were used and controls and schizophrenics were randomly and blindly measured. Thus, measurement artefact is not likely to account for the findings, as the authors used similar settings and conditions for both patients and controls. Head injury appears to be more associated with VE than does schizophrenia in this sample.

TWINS AND SINGLETONS COMPARED

While our findings on family history in schizophrenic twins were clear-cut, showing a clear distinction between familial and non-familial schizophrenia, so that all the twins with a family history of major psychiatric disorder have ventricular size below the control mean, these results were not consistently replicated in singletons. This suggested that twins might not be a good model for singletons. Inspection of the data also appeared to show that singletons obtained on the same scanner and measured in the same way, tended to have lower VBR and ventricular volume than twins. These results could not be explained on any methodological grounds, and therefore we developed the hypothesis that twins, as a group, had a larger ventricular size than singletons. Reasons for this might be that low birthweight and complicated delivery, are amongst the most important determinants of cerebral anoxia and haemorrhage in the newborn. As cerebral insults frequently give rise to ventriculomegaly, which may not be accompanied by neurological deficit, it suggested that twins who are particularly prone to such injury might therefore have mildly enlarged ventricles, even without neurological or behavioural deficit, as a result of adverse perinatal events or the complications of low birth weight. In that case, schizophrenic twins may have an even greater environmental component and therefore the familial-sporadic distinction would be easier to demonstrate in twins than singletons.

Accordingly, we [25] examined 25 schizophrenics of MZ twin birth, who were matched for age and severity of illness to 25 non twin schizophrenics who were current patients at the Maudsley Hospital. All were interviewed using SADS L and were diagnosed as having schizophrenia by both RDC and DSM III. When a first or second degree relative had had a psychiatric hospitalisation for major psychiatric disorder, excluding alcohol or drug abuse or had committed suicide, the family history was regarded as positive. Scans were obtained as previously described and VBR was calculated according to previous studies.[11]

The VBR for the 25 schizophrenic twins was 7.2 ± 4.7, while that for the comparison group of schizophrenic singletons was 5.0 ± 2.8 ($t=2.01$, $P<0.05$, 2-tailed). Analysis of variance of VBR by twin birth and family history controlling for age showed that age ($P=.004$), twin birth ($P=.038$) and family history ($P=.035$) were significant main effects predicting VBR. In addition, there was a significant ($P=.028$) twin x family history interaction. This was because when breaking the subjects down by family history, there was a much more striking difference in mean VBR between twins with a positive (4.15) vs negative (9.39) family history than between singletons with a positive (4.28) vs negative (5.77) family history.

Thus, a familial non-familial distinction resulted in a much clearer difference in VBR among the twin than among the singleton schizophrenics. However, there is a similar trend among the singleton schizophrenics. If these adverse environmental events were contributing to the development of schizophrenia in twins, one might expect a higher prevalence of schizophrenia among twins overall. However, prior studies have not found a higher prevalence of schizophrenia among twins than the general population rate.[26,27] These studies can be criticised on the basis of having an artificially healthy sample, inadequate diagnostic criteria, heterogeneity and failure to achieve complete ascertainment. Further, there is an increased probability that the co-twin of twins who later developed schizophrenia will die at birth, further artefactually reducing the prevalence of schizophrenia in twins. Thus, more careful studies might be able to demonstrate an excess of schizophrenia in twins, which would be environmentally mediated, making the familial sporadic distinction easier to detect.

REFERENCES

1. Reveley, MA (1985). CT scans in schizophrenia. Br J Psychiat, 146, 367-371.
2. Reveley, AM, Reveley, MA, Clifford, CA and Murray, RM (1982). Cerebral ventricular size in twins discordant for schizophrenia. Lancet, i, 540-541.
3. Spitzer RL, Endicott J and Robins E (1977) "The Schedule for Affective Disorders and Schizophrenia, Life-time version," 3rd edition. (New York:New York State Psychiatric Institute).
4. Spitzer RL, Endicott J and Robins E (1975). "Research Diagnostic Criteria Instrument No 58". (New York: New York State Psychiatric Institute.)
5. American Psychiatric Association (1980). "Diagnostic and Statistical Manual of Mental Disorders, 3rd edition." (Washington, DC:APA.
6. Reveley AM, Reveley MA and Murray RM (1984) Cerebral ventricular enlargement in non-genetic schizophrenia. Br J Psychiat, 144, 89-93.
7. Farr V (1975) Prognosis for the babies, early and late. In:Mac Gillivrary, P, Nylander PS and Corney G (eds). "Multiple Human Reproduction", pp 188-211.   (London:WB Saunders).
8.Thompson WD, Orvaschel H, Prusoff BA and Kidd KK (1982) An evaluation of the family history method for ascertaining psychiatric disorders. Arch Gen Psychiat, 39, 53-58.
9. Schulsinger F, Parnas J, Petersen ET, Schulsinger H, Teasdale TW, Mednick, SA, Moller, L, Silverton, L (1984). Cerebral ventricular size in the offspring of schizophrenic mothers:A preliminary study. Arch Gen Psychiat, 41, 602-606.
10. Reveley AM and Reveley MA (1984) Hypothesis for the aetiology of schizophrenia:A genetic perspective. Psychopharm Bull, 20, 505-508.
11. Reveley MA (1985) Ventricular enlargement in schizophrenia:validity of computerised tomographic findings. Br J Psychiat, 147, 233-240.
12. Reveley AM, Reveley MA, Chitkara B and Clifford C (1984) The genetic basis of cerebral ventricular volume. Psychiat. Res. 13, 261-266.
13. Falconer DS (1960). "Introduction to Quantitative Genetics". (London:Longman Group Limited).
14. Campbell, R, Hays, P, Russell,DB and Zacks DJ (1979). CT scan variants and genetic heterogeneity in schizophrenia. Am J Psychiat, 136, 722-723.
15.Pearlson, GD, Garbacz, DJ, Moberg, PJ, Ahn HS and De Paulo JR (1985). Symptomatic, familial, perinatal and social correlates of computerised axial tomography (CAT) changes in schizophrenics and bipolars. J Nerv Ment Dis, 173, 42-50.
16. Farmer, A, McGuffin, P, Jackson, R and Storey P (1985). Classifying schizophrenia. Lancet, i, 1333.
17. Nasrallah, HA, Kuperman, S, Hamra, BJ and McCalley-Whitters, M. (1983). Clinical differences between schizophrenic patients with and without large cerebral ventricles. J Clin Psychiat, 44, 407-409.
18. Oxiensterna, G, Bergstrand, G, Bjerkenstedt, L, Sedvall, G and Wik, G (1984). Evidence of disturbed CSF circulation and brain atrophy in cases of schizophrenic psychoses. Br J Psych, 144, 654-

661.

19. Reveley, MA and Chitkara, B (1985). Subgroups in schizophrenia. Lancet, i, 1503.

20. Cazzullo, CL, Sacchetti, E, Vita, A, Illeni, MT, Bellodi, L, Maffei, C, Alciati, A, Bertrando, P, Calzeroni, A, Ciussani, S, Conte, G, Pennati, A and Invernizzi, G (1985). Cerebral ventricular size in schizophrenia spectrum disorders: relationship to clinical, neuropsychological, and immunogenetic variables. Abstracts 4th WCBP, Philadelphia, 7-11 Sept, p. .

21. Turner, SW, Toone, BK and Brett-Jones, JR (1986). Computerised tomographic scan changes in early schizophrenia: Preliminary findings. Psychol Med, 16, 219-225.

22. Owens, DGC, Johnstone, E, Crow, TJ et al (1985) Lateral ventricular size in schizophrenia: Relationship to the disease process and its clinical manifestations. Psychol Med, 15, 27-41.

23. Weinberger GR, DeLisi LE, Naophytides AN and Wyatt RJ (1981) Familial aspects of CT scan abnormalities in chronic schizophrenic patients. Psychiat Res, 4, 65-71.

24. DeLisi LE, Goldin LR, Hamovit JR, Maxwell E, Kurtz D and Gershon ES (1986) A family study of the association of increased ventricular size with schizophrenia. Arch Gen Psychiatry, 43, 148-153.

25. Reveley AM and Reveley MA (1987) The relationship of twinning to the familial-sporadic distinction in schizophrenia. J Psychiat Res, in the press.

26. Allen MG and Pollin W (1970) Schizophrenia in twins and the diffuse ego boundary hypothesis. Amer J Psychiat, 127, 437-442.

27. Rosenthal, D (1960). Confusion of identity and the frequency of schizophrenia in twins. Arch Gen Psych, 3, 297-304.

11
CT scans of the brain in families with a highfrequency of schizophrenia

G. Niu-Fan, Y. He-Qin, C. Guo-Jung,
Z. Ming-Dao, Z. Ming-Yuan and
X. Zhen-Yi

This study was based on the examination of the CT scan changes in a group of schizophrenic patients from families with a high frequency of schizophrenia(HFS). It is our hypothesis that the patients composed a relatively homogenous sub-group. We chose to compare the CT scan measures of our schizophrenic patients with two other groups, a normal first degree relative group and a normal control group. Our purpose was to look for specific abnormalities in the brains of these three groups in order to evaluate their relationship to clinical manifestations of illness, prognosis and morbidity risk.

METHODS
Subjects

All patients fulfilled the diagnostic criteria of DSM-III and the Yellow Mountain Conference guidelines (China,1984) and were in-patients and out-patients of the Shanghai Mental Health Center. A HFS of schizophrenic patient is defined as one coming from a family having had two or more schizophrenic patients in two successive generations. The Brief Psychiatric Rating Scale(BPRS) was administered by G.J.Cai before the CT scans. Exclusion criteria for possible subjects were 1) a diagnosis of any CNS disease; 2) any clinical or laboratory evidence of marked dehydration, volume depletion, or severe weight loss; 3) treatment with systemic steroid or radiation thrapy; 4) a history of any other psychiatric diagnosis; 5) a history of drug or alcohol abuse.

There were 24 patients from 9 high frequency families,14 males and 10 females with a mean age of 38.5±13.3 years(19-63 years), and a mean illness duration of 11.3±8.8 years(2-29 years).In the 9 families, 16 first degree relatives were over 17 years, including 7 males, 9 females, with an average age of 37.3±14.3 years(17-67 years).

The control group was composed of 40 hospital staff members and medical students matched with patients and their relatives for sex and age(±2 years). The exclusion criteria for controls were the same as those for patients.

Mental Status

For the most part, the patients were apathetic and had autistic thinking and emotional disturbances. Some were delusional and mal-adapted. Of the total number of patients, 8 attended normal schools or worked, 6 did not receive any medication while 18 were under antipsychotic drug therapy with an average dosage equal to CPZ 220 mg/day.

CT Scans and Measurement

All CT scans were performed with a ▲-100 CT machine(made in the USA). The measurements(1) were made independently by at least two raters, who were not aware of the subjects' mental status. In order to avoid subjective error in the measurement, the raters measured each subject's scan 5 different times and then calculation their mean values.

RESULTS AND ANALYSIS

CT Findings

Table 1 shows that the sizes of both the ventricle brain ratio(VBR) and 3rd ventricle of the schizophrenic patients were the largest,those of their first degree relatives of middle range and those of the control group smallest. The differences in the width of the 3rd ventricle in these tree groups were statistically significant(F-test,$P<0.05$). In order to compare the width of the 3rd ventricles of schizophrenic patients with the other two groups, we performed a Q-test. The results showed a significant difference between the schizophrenic and the control group.But no difference betweenthe schizophrenics and their first degree relatives. Therefore, we combined the schizophrenic patient and first degree relative groups.A significant difference was found between this combined group and the controls($P<0.05$).

The values of the brain density of the frontal and parietal lobes of the first degree relatives and normal controls were higher than those of the schizophrenics.The brain density of the temporal lobe and cerebellum of schizophrenic patients was higher than in the other groups, but these differences were not statistically significant.

The value of more than 1.96 SD beyond the control mean was used as a cut-off point for defining the abnormality of the 3rd ventricle or VBR. The value of either the 3rd ventricle or VBR in nine of the schizophrenic patients and four first degree relatives was above these cut-off point.One schizophrenic patient had both 3rd ventricle size and VBR abnormalities. A significant difference was found in the frequency of abnormality of the 3rd ventricle and VBR when the HFS subjects and controls were compared(Table 2, $P<0.05$).

The Width of the 3rd Ventricle Versus Other CT Findings

No significant correlations were noted when VBR, brain density and the width of 3rd ventricle were compared.

Table 1. VBR and 3rd ventricle results

	Schizo. group N=24	I°relative N=16	Control group N=40	F value	P value
VBR	7.18±3.37	6.92±2.99	6.09±2.65	1.15	>0.05
Width of 3rd ventricle(mm)	4.50±1.30	4.20±1.21	3.67±1.04	4.08	<0.05

Table 2. Changes of VBR and 3rd ventricle in Schizophrenics and their I° relative compared with control group*

Abnormality (case,%)	Schizo. group N=24	I°relative N=16	Control group N=40	X^{2**}	P value
VBR	4(16.70)	2(12.20)	0	5.64	<0.05
3rd ventricle	5(20.80)	2(12.20)	0	4.50	<0.05
Total	9(37.50)	4(25.00)	0	15.52	<0.05

*Abnormality: Mean value(control group)+1.96 SD,i.e. 3rd ventricle> 5.71mm, VBR>11.28
**X^2 value calculated with corrected X^2 test, includes schizophrenic patients and I° relative combined

Table 3. Duration of illness in relation to 3rd ventricle and VBR

	>5 years N=16	<5 years N=8	t value	P value
Width of 3rd ventricle (x±SD)	4.75±1.25	4.00±1.35	1.35	>0.05
VBR (x±SD)	8.16±3.18	5.22±3.02	2.17	<0.05

Clinical Manifestations Versus CT Findings

The 24 schizophrenic patients were divided into two sub-groups: the paranoids(11 cases) and non-paranoids(13 cases). The CT findings for these three groups did not show any significant differences between them. Further, no significant correlation was found between the width of 3rd ventricle and the duration of illness;however,the size of VBR tended to increase following prolonged ilness.The VBR value was greater in patients who were ill for over 5 years than in those who had been sick for less than 5 years(Table 3.t=2.17,P<0.05).

CT Findings Versus ECT and Prognosis

Of the 9 schizophrenic patients who had received ECT(6-12 sessions), 4 showed larged 3rd ventricle(>5.71mm) and 2 showed large VBR(>11.28).

Of the 13 schizophrenic patients who had not received ECT,2 showed large VBR. No statistical significance was found between these two groups.No statistical significance was found between the size of the 3rd ventricle and VBR when well-adapted and mal-adapted patients were compared.

CT Findings and BPRS Score

After having analyzed the BPRS factors involving schizophrenic symptoms,no significant differences between these factors and the CT findings were found.

Clinical Genetics Versus CT Findings

We divided HFSs into two sub-groups: schizophrenic father group(F-group,25 cases) and schizophrenic mother group(M-group,15 cases).We found that the sizes of the 3rd ventricles of the M-group were larger than those of the F-group, 4.74±1.15 mm and 3.76±1.23 mm,respectively(Table 4, P<0.05).

We further divided all the HFSs into 2 groups on the basis of the patients' paranoid symptoms: paranoid group of father/mother(13 cases) and non-paranoid group of father/mother(27 cases).Here, the sizes of VBR of the non-paranoid group were greater than those of the paranoid group, 8.68± 3.30 and 6.30± 2.88,respectively. There was significant difference between the two groups(Table 5, P<0.05).

DISCUSSION

A great many articles on the CT scans of schizophrenic patients have been published(2-6). Woods and Wolf(7) indicated that many factors might be involved in the results of these CT scans. Until short time ago,considerable attention was given to differences in radiological methods as a means of explaining different findings in different studies. On the contrary, relatively little weight was given to potential differences in patient populations and control groups.Consequently, we assembled all the HFS subjects,including the schizophrenic patients and their normal first degree relatives,to serve as a group of "homogenous" subjects to be compared with normal controls. This

Table 4. CT findings in schizophrenic parents

	schizophrenic father N=15	schizophrenic mother N=25	P value
Width of 3rd ventricle (x±SD)	3.76±1.23	4.74±1.15	<0.05
VBR (x±SD)	7.42±3.57	6.86±2.99	>0.05
Abnormality of 3rd ventricle.case(%)	1 (6.7)	6 (24.0)	>0.05
Abnormality of VBR. case(%)	4 (26.7)	2 (8.0)	>0.05

Table 5. CT findings in paranoid and non-paranoid schizophrenics

	father/mother paranoid N=13	father/mother non-paranoid N=27	P value
Width of 3rd ventricle (x±SD)	4.28±1.44	4.58±0.79	>0.05
VBR (x±SD)	6.30±2.88	8.68±3.30	<0.05
Abnormality of 3rd ventricle.case(%)	2 (15.4)	5 (18.5)	>0.05
Abnormality of VBR. case(%)	3 (23.1)	3 (11.1)	>0.05

was doing as a possible means of making the morphological measurements more reliable.

Our findings indicated that the sizes of the 3rd ventricle and VBR of the schizophrenic patients and their relatives were greater than those of the control group. The 3rd ventricles or VBR in 37%(9/24 cases) of the schizophrenic patients and 25%(4/16 cases) of their first degree relatives were abnormal. These morphological changes were not secondary to hospitalization or treatment since we also found the same characteristics in the patient' first degree relatives.

Tanaka et al.(8), Okasha and Madkour(9) and Boronow et al.(10) did not find lateral cerebral ventricular enlargement(11-12),but did mention that schizophrenic patients had significantly large 3rd ventricles. Nyback et al.(13) stated that both schizophrenic patient and control group subjects had 3rd ventricle enlargement. Recently Nasrallah et al.(14) reported that schizophrenic patients had 3rd ventricle enlargement,which is associated with cerebellar atrophy.

Stevens(15) found that neurogliosis surrounding the 3rd ventricle might cause cerebral atrophy. Rosenthal and Bigelow(16) and Bigelow et al.(17) reported that the corpus callosum was thick in chronic schizophrenic patients. As we know,the limbic system and the hypothalamus,with amine- or peptide-containing neurons,are close to the 3rd ventricle.Enlargement of the 3rd ventricle thus may be due to atrophy of these hypothalamic nuclei. The destruction of this nerve tissue containing neuropeptides may give rise to hypothalamic dysfunction and affect mental activity.The dilation of the 3rd ventricle might be an organic vulnerability in some schizophrenic patients.If some other factors,such as metabolic disturbances,chronic malnutrition, chronic infection or psychosocial stress are imposed on patients with this organic vulnerability,their schizophrenic symptoms may then be more pronounced.The same explanation may hold for the first degree relatives of such patients.

From the clinical genetics view point,we found that the size of the 3rd ventricle was larger in most family members with a schizophrenic mother,but not in family members with a schizophrenic father. Thus,theorganic vulnerability of schizophrenic patients—enlargement of the 3rd ventricle—might be more readily transmitted to the offsprings by their mother.

King et al.(18) reported that the VBR of paranoid schizophrenia was within the normal range.With our retrospective study we found that the personlity of the schizophrenic patient with a paranoid parent is integrated.Our findings suggest that schizophrenic patients with a normal VBR have a tendency to greater personality integration.

In the future,we should extend the size of HFS of schizophrenic patient samples and combine the morphological research with physiological,biochemical and pharmacokinetic variables as well as undertake longitudinal follow-up studies.

ACKNOWLEDGEMENTS
We wish to express our gratitude to Zhen-Yiu Wang, Ping Fan, and Yiu-Hua Tang for their computer programming assistence, to Yue-Fen Zhie for her collaboration, and to Professor Leiv R. Gjessing and Professor E. Sacchetti for reviewing this manuscript.

REFERENCES

1. Synek,V and Reuben,JK (1976). The ventricular-brain ratio using plantimetric measurement of EMI scans. Brit J of Radiology 49,233
2. Schulz,SG et al (1983). Ventricular enlargement in teenage patients with schizophrenia spectrum disorder.Am J Psychiatry 140,1592
3. Tsai, LY et al (1983).Hemispheric asymmetries on computed tomographic scans in schizophrenia and mania.Arch Gen Psychiatry 40,1286
4. Largern, JW et al (1984). Abnormalities of brain structure and density in schizophrenia. Biol Psychiatry 19,991
5. Luchins, DJ et al (1984).Lateral ventricular size,psychopathology, and medication response in the psychoses. Biol Psychiatry 19, 29
6. Schulsinger,F et al (1984).Cerebral ventricular size in the offspring of schizophrenic mother(A preliminary study).Arch Gen Psychiatry 41,602
7. Woods,BT and Wolf,J (1983). A reconsideration of the relation of ventricular enlargement to duration of illness in schizophrenia. Am J Psychiatry 140,1564
8. Tanaka,Y et al (1981). Computerized tomography of the brain in schizophrenic patients. Acta Psychiat Scand 63,191
9. Okasha, A and Madkour,O (1982). Cortical and central atrophy in chronic schizophrenia. Acta Psychiat Scand 65,29
10. Boronow J et al (1985). Atrophy limited to the third ventricle in chronic schizophrenic patients. Arch Gen Psychiatry 42,266
11. Weinberger DR et al (1979). Lateral cerebral ventricular enlargement in chronic schizophrenia. Arch Gen Psychiatry 36,735
12. Weinberger DR et al (1979). Structural abnormalities in cerebral cortex of chronic schizophrenic patients. Arch Gen Psychiatry 36,935
13. Nyback,H et al(1982). Computed tomography of the brain in patients with acute psychoses and in healthy volunteers.Acta Psychiat Scand 65,403
14. Nasrallah,HA et al(1985). Third ventricular enlargement on CT scans in schizophrenia:Association with cerebellar atrophy. Biol Psychiatry 20,443
15. Stevens, JR (1982). Neuropathology of schizophrenia. Arch Gen Psychiatry 39,1131
16. Rosenthal,R and Bigelow,LB (1972). Quantitative brain measurements in chronic schizophrenia. Brit J Psychiatry 121,259
17. Bigelow, LB et al (1983). Corpus callosum thickness in chronic schizophrenia. Brit J Psychiatry 142,284
18. King, AS et al (1982/83). CT scans in sub-groups of chronic schizophrenics. J Psychiat Res 17,375

12
Magnetic resonance imaging: applications in psychiatry

N.C. Andreasen, H.A. Nasrallah,
J.C. Ehrhardt, W.M. Grove,
S.C. Olson, J.A. Coffman and
J.H.W. Crossett

Magnetic resonance imaging (MRI) is a relatively new radiological technique that is especially useful for evaluating brain structure. It offers several major advantages over computerized tomography (CT), the technique that has been most widely used in psychiatry to date and that has provided clear evidence for structural brain abnormalities in schizophrenia [1-3]. Unlike CT, MRI does not require the use of ionizing radiation. It permits visualization of the brain in multiple planes including coronal, sagittal, and transverse. It gives impressive gray-white resolution, yielding pictures nearly as anatomically precise as can be seen directly in postmortem brains. It is also highly sensitive to detecting small white matter lesions. It is not subject to bony artifacts and thereby permits excellent visualization of structures in the posterior fossa.

We have applied this technique to a sample of 38 patients suffering from schizophrenia, who were compared with 49 normal controls. This study had several facets, but one primary goal was to search for structural brain abnormalities that are more focal than can be visualized on CT. Specifically, we were interested in determining whether patients suffering from schizophrenia exhibit structural abnormalities in their frontal lobes.

Injury to the frontal lobes through trauma or disease produces a broad range of deficits in higher emotional and behavioral functions, including avolition, impaired attention, difficulty in generating fluent spontaneous speech, and abnormalities in affect [4]. These symptoms are quite similar to those noted in patients suffering from schizophrenia, particularly those with prominent negative symptoms. Further, studies of cerebral blood flow and brain electrical activity have also suggested that abnormalities in the frontal system may occur in schizophrenia [5-7]. Given this constellation of evidence from both the clinical picture of the illness and metabolic-functional studies, it seemed logical to determine whether any underlying structural abnormalities could be observed.

Our preliminary studies have been limited to measurement of a single midline sagittal cut. A photograph of this cut appears in Figure 1. As this figure shows, sulci can be seen clearly and structures well delineated, including frontal, parietal, and occipital lobes, cingulate gyrus, corpus callosum, etc. In this preliminary work we were able to obtain good interrater reliability for three major regions: cerebral area (Pearson r = .97); frontal area (Pearson r = .81); and cranial area (Pearson r = .91). Analyses reported herein are limited to these structures.

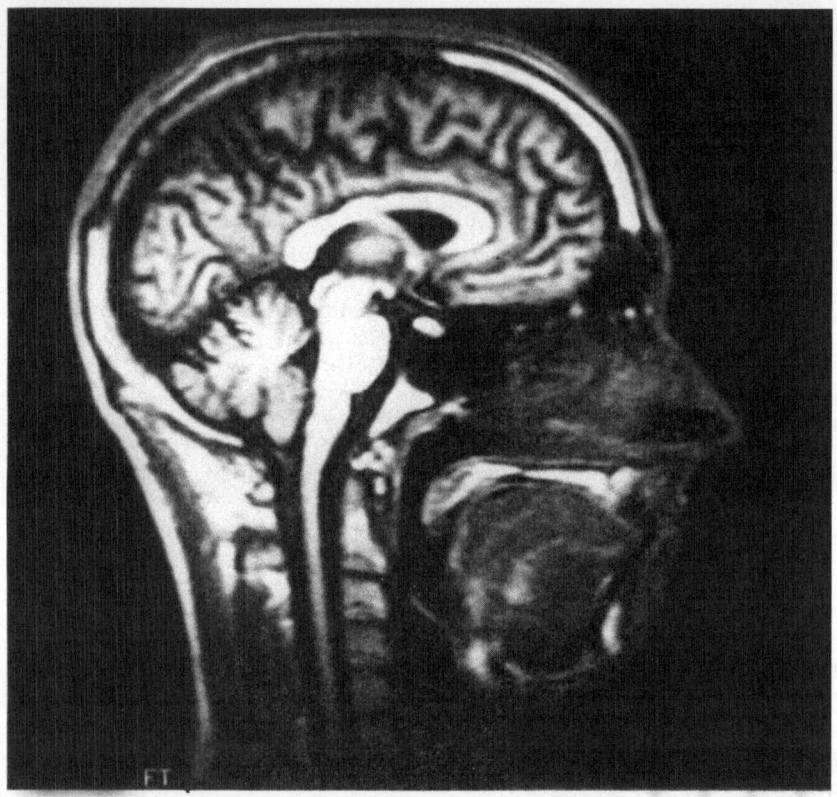

FIGURE 1. Midline sagittal cut using MRI

The 38 patients suffering from schizophrenia were drawn from consecutive admissions to the University of Iowa Psychiatric Hospital (N = 20) or the Iowa City Veterans Hospital (N = 18). Their mean age was 33.18 ± 8.16; their mean age at onset was 22.01 ± 4.17; their mean duration of hospitalization was 10.83 ± 10.36; and their mean number of hospitalizations was 6.86 ± 5.37. The control subjects consisted of 49 healthy volunteers recruited from hospital personnel. They were slightly younger than the subjects with a mean age of 27.65 ± 4.89. Neither patients nor controls were taking medications known to affect brain size such as steroids. The patients and controls did not differ in height and weight.

All subjects were examined using a Picker MRI system with a .5 Tesla superconducting magnet. The scanning sequence consisted of a T1 weighted image (inversion recovery with a typical inversion time of 600 msec and a typical repetition time of 1,600 msec). Scans were measured blindly. A calibration line on each scan permitted measurement of actual brain size in each region in square centimeters.

Table 1 shows the results of measuring frontal, cerebral, and cranial areas for both male and female subjects. Stratification by sex is necessary because the samples are not balanced and because of sex differences in head and brain size.

TABLE 1. Frontal, cerebral, and cranial area measured on midsagittal cut in schizophrenics vs. controls*

Area and Sex	Schizophrenics		Controls	
	N	Mean ± SD	N	Mean ± SD
Frontal area				
All subjects	38	45.65 ± 6.09	49	49.25 ± 5.99
M	28	46.03 ± 6.80	25	51.98 ± 5.42
F	9	44.48 ± 2.95	24	46.41 ± 5.28
Cerebral area				
All subjects	38	86.25 ± 10.70	49	91.18 ± 10.63
M	28	86.47 ± 11.38	25	96.31 ± 9.38
F	9	85.57 ± 8.83	24	85.85 ± 9.29
Cranial area				
All subjects	38	159.91 ± 12.80	49	164.15 ± 14.69
M	28	159.22 ± 11.18	25	171.60 ± 13.05
F	10	161.85 ± 17.11	24	156.39 ± 12.22

* Measurements are actual areas in square centimeters; frontal and cerebral data are missing for one subject because of poor quality due to movement.

Table 2 examines these data statistically through analysis of variance. For frontal size there is a significant diagnostic and sex effect and a trend toward diagnosis by sex interaction. For cerebral size there is a prominent diagnostic effect as well as a significant sex effect and a significant diagnosis by sex interaction. For cranial size there is a trend toward a diagnostic effect and a strong sex effect and diagnosis by sex interaction. As shown in Table 1, these differences arose primarily because the male schizophrenics had smaller frontal, cerebral, and cranial areas.

119

TABLE 2. Analysis of variance of diagnosis and
sex in frontal, cerebral, and cranial
areas measured on midsagittal cut

Area	Source	F	p
Frontal	Diagnosis	8.41	.005
	Sex	9.99	.002
	Diagnosis by sex	2.16	.145
Cerebral	Diagnosis	5.12	.026
	Sex	9.44	.003
	Diagnosis by sex	3.99	.049
Cranial	Diagnosis	2.35	.129
	Sex	8.70	.004
	Diagnosis by sex	8.97	.004

We also determined how many patients differed markedly
from the controls by determining whether they were outside
the control range or more than two standard deviations below
the control mean. Eleven patients (39%) were outside the
control range on frontal size and also more than two standard
deviations below the control mean. Seven (25%) were below
the control range on cerebral size and six (21%) were more
than two standard deviations below the control mean. Five
(18%) patients were below the control range in cranial size
and five (18%) more than two standard deviations below the
control mean. The analysis of variance seemed to suggest that
the difference between the schizophrenics and controls was
contributed by the male subjects. Nevertheless, it is
premature to draw that conclusion at this point.

We used multiple regression to remove the effects of sex,
height, and weight. When this was done, these results
suggested that some female schizophrenics also had
disproportionately small cranial, cerebral, and frontal lobe
size. Consequently, a second larger sample is needed in order
to determine more accurately the extent to which these
findings may be sex specific. These analyses also documented
that the differences between patients and controls were not
due to artifacts such as body size, but rather that the
patients had disproportionately small cranial, cerebral, and
frontal lobe size.

It is important to determine whether the decrease in
cerebral size is due to a diffuse decrease or whether it is due
to a specific decrease in frontal lobe size. Analysis of
covariance was used in order to determine this. The results
of this analysis indicated that the decrease in frontal lobe
size is a specific finding and not secondary to a decrease in
either cranial or cerebral size.

The decreased frontal lobe size had been predicted a priori, as had the decrease in cerebral size. The finding of decreased cranial size was unexpected, however. This finding suggests the possibility that we must change our perspective concerning the pathological process that occurs in some patients suffering from schizophrenia. Heretofore, the structural brain abnormalities observed in schizophrenia have been referred to and thought of as possibly atrophic. The finding of small cranial size suggests that the process could instead be dystrophic. In young children cranial growth occurs in harmony with brain growth. Cranial growth is essentially complete within the first two years of life; after that time the sutures close and cranial growth can only occur through the relatively slow process of modelling. The small cranial size observed in this study suggests that the pathological processes in schizophrenia may begin early and result from a variety of environmental influences, such as intrauterine nourishment, perinatal complications, or early postnatal influences such as infections or head injuries. These findings have been partially confirmed by several other recent studies. Benes et al. have reported a decrease in both neuronal and glial density in the prefrontal cortex of schizophrenics, a finding suggestive of early developmental mechanisms rather than a degenerative process [8].

Schulsinger et al. have observed a relationship between increased ventricular size and low birth weight, a finding that also suggests that the pathological process occurring in schizophrenia may begin relatively early [9]. The Northwick Park group has found patients suffering from schizophrenia to have decreased brain weight as measured at postmortem [10].

The study of the brain with MRI is just beginning. This technique promises to have many applications in psychiatry. As the present study suggests, it can be used to determine whether structural abnormalities occur in some patients suffering from schizophrenia. The present investigation is limited to the midsagittal cut. Many structures can be seen on coronal cuts that are also of potential significance to the phenomenology of schizophrenia, such as the temporal lobes, the amygdala, the hippocampus, or the basal ganglia.

Ultimately, as this study suggests, brain imaging techniques such as MRI can be used to illuminate the mechanism behind observed neural abnormalities in schizophrenia. The results of this particular study suggest that these abnormalities may represent a dystrophic or developmental process, which affects principally the frontal lobes, rather than an atrophic or degenerative one.

REFERENCES

1. Johnstone, E.C., Crow, TJ, Frith, CD, Husband, J, and Kreel, L (1976). Cerebral ventricular size and cognitive impairment in chronic schizophrenia. Lancet, 2, 925.
2. Weinberger, DR, Torrey, EF, Neophytides, AN, and Wyatt, RJ (1979). Lateral cerebral ventricular enlargement in chronic schizophrenia. Arch Gen Psychiatry, 37, 735.
3. Andreasen, NC, Smith, MR, Jacoby, CG, Dennert, JW, and Olsen, SA (1982). Ventricular enlargement in schizophrenia: Definition and prevalence. Am J Psychiatry, 139, 292.
4. Fuster, JM (1980). The Prefrontal Cortex. (New York: Raven Press).
5. Weinberger, DR, Berman, KF, and Zec, RF (1986). Physiological dysfunction of dorsolateral prefrontal cortex in schizophrenia: 1. Regional cerebral blood flow evidence. Arch Gen Psychiatry, 43, 114.
6. Buchsbaum, MS, Ingvar, DH, Kessler, R, Waters, PN, Cappelletti, J, Van Kammen, DP, King, AC, Johnson, JL, Manning, RG, Flynn, RW, Mann, LS, Bunney, WE, and Sokoloff, L (1982). Cerebral glucography with positron tomography: Use in normal subjects and in patients with schizophrenia. Arch Gen Psychiatry, 39, 251.
7. Morihisa, JM, Duffy, JH, and Wyatt, RJ (1983). Brain electrical activity mapping in schizophrenic patients. Arch Gen Psychiatry, 40, 719.
8. Benes, FM, Davidson, J, and Bird, ED (1986). Quantitative psychoarchitectural studies of the cerebral cortex in schizophrenics. Arch Gen Psychiatry, 43, 31.
9. Schulsinger, F, Parnis, J, Peterson, ET, Schulsinger, H, Teasdale, TW, Mednick, SA, Moller, L, and Silverton, L (1984). Cerebral ventricular size in the offspring of schizophrenic mothers: A preliminary study. Arch Gen Psychiatry, 41, 601.
10. Brown, R, Colter, N, Corsellis, JAN, Crow, TJ, Frith, CD, Jagoe, R, Johnstone, EC, and Marsh, L (1986). Postmortem evidence of structural brain changes in schizophrenia. Arch Gen Psychiatry, 43, 36.

13
Neurofunctional assessment of schizophrenia: neuropsychological and EEG data of lateralized imbalancement of functions

S. Scarone, O. Gambini, L. Pugnetti,
C. Colombo, R. Cattaneo, L. Bellini,
W. Christe and C.L. Cazzullo

INTRODUCTION

The purpose of this paper is to discuss how our data on hemispheric specialization measures might be usefull for improving our understanding of schizophrenia.

In the last ten years most findings from many independent studies have substantiated the hypothesis that in neurofunctional terms, the schizophrenic syndrome is characterized by an impairment in the organization of lateralized hemispheric functioning. This viewpoint has had its theoretical basis in the well known neurophysiological organization of the Central Nervous System, in which the thinking processes are attributed to the dominant hemisphere- the left one in right handed subjects. The non-dominant hemisphere- the right one in right handed people- sustains the emotional and affective processes.

Initially, most of the research in this area was devoted to delineating the main characteristics of hemispheric functioning in schizophrenic patients as compared to normal controls or to patients affected by other psychiatric disorders.

The results obtained (Tab 1) would indicate that the schizophrenic syndrome is asssociated with a high degree of abnormal hemispheric functioning, primarly in the left hemisphere. In the more recent past, some reports have focused on the possible relationship between the clinical and epidemiological characteristics of schizophrenia and different degrees of hemispheric malfunctioning. Table 2 summarizes the main results of this approach. Consistent with other evidences, lateral process impairment also appears to be dependent on certain clinical and epidemiological characteristics though not with 100% reliability.

Table 1. Hemispheric malfunctioning in schizophrenia: Major
findings.

Left temporal EEG abnormalities and left fronto-temporal increase
in fast activity in adult patients (1,2)

More slow activity in left frontal regions of chronic patients
(3,4)

Reversal of normal asymmetries in chronic patients (5)

Impairment of dominant temporo-parietal neuropsychological
performances (6)

Left hemisphere impairment in visual and verbal information processing (7)

Left side autonomic hypoactivity in chronic patients (8)

Consequently, it is plausible to suppose that, like other biological characteristics of the disease, the neurofunctional organization of the C.N.S. in schizophrenia might be dependent on
patients traits and/or states characteristics.
On the other hand, some phenomena that are no so strictly
biological- psychopharmacological treatment, number and duration
of hopsitalizations, social and familial assessment of individual
patients- seem to play a critical role in determining the degree
of lateral malfunctioning in schizophrenia.
This paper will discuss some of our more recent neuropsychological and neurophysiological data in this field.

Neuropsychological Studies
    In a recent series of publications the present investigators
reported that schizophrenic patients, like neurological patients
with contralateral parietal or homolateral frontal lesions
(14), had a higher incidence of left- sided extinctions on
Schwartz's Quality Extinction Test (QET) (15, 16). Taking the
organization of the central tactile sensitivity pathways into
account, the QET is able to discriminate between specific parietal and/or frontal area malfunctioning and more general unspeci-
fic central area damage due to a subcortical attention mechanism.
Tactile extinction has been found to be independent of patients'
nosographic classification (Tab. 3) but dependent on disease

124

course (17) (Tab.4).

Table 2. Hemispheric malfunctioning in schizophrenia: clinical correlates

---

Higher alpha power of the left hemisphere in hebephrenics and lower alpha power in paranoids (9)

Left hemisphere abnormalities in florid patients, right hemisphere abnormalities in retarded catatonics (10)

Left hemisphere dysfunctions in delusional patients, bilateral malfunctioning in "affective" patients (11)

Non-dominant frontal dysfunction in acute patients: pathological activation of dominant temporo-frontal regions in chronic patients (1)

Abnormal increase of fast beta activity in non-dominant hemisphere in young patients: increase of fast beta activity in the dominant hemisphere in chronic patients (12).

Functional impairment of non-dominant parietal regions and/or dominant frontal regions in chronic patients (13)

Table 3. Distribution of tactile extinctions according to diagnosis in schizophrenic patients

---

	Extinctions		
	Left	Right	None
Diagnosis			
Disorganized	27	7	24
Paranoid	12	1	22
Undifferentiated	20	3	20

Table 4. Distribution of tactile extinctions according to the chronicity of schizophrenia

	Extinctions		
	Left	Right	None
Chronicity (Total sample)*			
Subchronic	5	4	15
Chronic	56	7	49
Chronicity (Age-matched)**			
Subchronic	5	4	15
Chronic	15	2	7

* chi-square = 7.88, d.f.=2, P < 0.025
**chi-square = 8.58, d.f.=2, P < 0.025

In general, findings to date revealed that, compared to normal controls, schizophrenic patients had a significantly higher incidence of left extinguishing responses to the QET and a greater number of left and right extinctions. With regard to the latter, the left extinctions were significantly more frequent than the right ones (16). It therefore, seems that QET responses in schizophrenia are independent of the clinical aspects of the illness and instead related to some type of neurofunctional impairment in hemispheric functioning inherent in the disease itself. Other variables, such as duration and number of hospital admissions or duration of psychopharmacological treatment might be important in determining QET responses in schizophrenia. But only continued study will allow greater understanding of tactile sensitivity and its precise relationship to schizophrenia.

In any case, it is apparent fact that a better understanding of the close relationship between the biological characteristics of schizophrenia and patients' clinical and epidemiological traits might assist in identifying diverse limited group of patients with similar biological traits, disease course and prognosis and therefore give us a useful basis for formulating preventive strategies.

Table 4. Clinical and demographic characteristics

Diagnostic Category	N	Sex	Age	Duration of the illness (ys)
Normal Controls	9	6M,2F	25.2(3.2)	–
Schizotypal and Schizoid Personality Dis	9	5M,3F	21.7(3.4)	–
Subchronic Schizophrenia Disorganized Type	9	6M,2F	21.1(3.4)	1.5 (0.5)
Chronic Schizophrenia Disorganized Type	9	6M,2F	22.4(3.7)	5.7 (1.7)

Table 5. Summary of the Analysis of Variance

FREQUENCIES   FACTORS

	DIAGNOSTIC CATEGORIES	COGNITIVE TASKS	RECORDING SIDE
DELTA	$F(3,32)= 3.63**$	$F(2,64)= 34.19***$	
THETA	$F(3,32)= 4.21**$	$F(2,64)= 11.31***$	
THETA	$F(3,32)= 3.85**$		
ALPHA		$F(2,64)= 11.90***$	
ALPHA2	$F(3,32)= 4.61**$	$F(2,64)=\ 6.30***$	
BETA1			
BETA2		$F(2,64)=\ 6.21***$	

INTERACTIONS

	DIAGNOSTIC CATEGORIES x COGNITIVE TASKS	DIAGNOSTIC CATEGORIES x RECORDING SIDE
DELTA		
THETA1	$F(6,64)= 3.18***$	
THETA2		
ALPHA1	$F(6,64)= 2.41*$	
ALPHA2		
BETA1	$F(6,64)=\ 4.54***$	
BETA2	$F(6,64)= 4.52***$	$F(3,32)= 3.30*$

* $p < .05$
** $p < .02$
*** $p < .01$

EEG studies

General EEG characteristics. Previous research consistently has pointed to a high incidence of EEG abnormalities in schizophrenia. Initially, visual assessment yielded several non-specific findings like poor alpha rythm , increased beta activity and disorganized activity (18) . The introduction of new computerized techniques now makes it possible to evaluate the electrical cerebral background in a more sophisticated way (19). Consequently, the hypovariability of EEG activity (20), an excess of beta activity (21) and decreased alpha power (22) also have been reported.

Discrepancies in the results from different laboratories are likely due to factors like varying technical and sampling procedures.

EEG Hemispheric Characteristics. Several researchers have reported unusual but inconsistent findings of lateralized EEG patterns. In fact, left side malfunctioning(23), right side dysfunction (24) or no abnormalities(25) have been described. Again, these discrepancies might be artifacts of differences in the hemispheric activation and/or sampling procedure used by different investigators.

Diagnostic and Psychopathological Characteristics. The literature presents disparate and frequently contrary findings. Hebephrenic and paranoid patients are different (26) or not different (2) from each other in terms of hemispheric malfunctioning. Florid patients seem to show a predominant left hemisphere pathology(27) but this also is manifested in chronic undifferentiated patients who have no positive symptomatology (26).
Recently, we assessed the EEG hemispheric characteristics (recorded from P3-Cz and P4-Cz under resting conditions and during two different cognitive tasks) of three different patient groups, one composed of subjects with a diagnosis of schizotypal or schizoid personality disorder and a second of patients with subchronic schizophrenia who had never received systematic psychopharmacological treatment.
Third, we selected a group of chronic young patients with an illness duration between 2 and 7 years who had not received drug treatment for at least one month prior to the study.

Tab 4 shows the clinical and epidemiological characteristics of the sample.

Tab 5 show the summary of the analysis of variance. As compared with controls, subchronic patients show a significant excess of slow activity in delta and thetal frequency bands; chronic patients, personality disorders and controls do not differ from each other. Subchronic patients also show a significant reduction in alpha2 activity once again, there are no statistical differences between the other three groups. The performance of cognitive task affects the power of almost all the frequency bands, with an increase in slow and fast and a decrease of alpha activity, without any significant effect on the recording side. Diagnosis was significant for the recording side of fast beta lateralized distribution power.When compared to controls, the three groups of patients showed a significant excess of fast beta activity on the left hemisphere whereas the personality disorder patients and subchronic patients had a significant excess of fast beta on the right hemisphere as well. There were no significant interhemispheric differences among controls or patient groups.

Effects of motor and cognitive tasks. The effects of motor and cognitive tasks on lateralized EEG characteristics are still far from consistent. Tab. 6 shows the main recent findings on the problem. In particular, there is no univocal evidence for task-related hemispheric malfunctioning, even if a left-sided dysfunction seems to be prevalent.

Effects of psychotropic drugs. As a further complication there is the stil controversial influence of psychotropic agents on the EEG indices of hemispheric functioning.

Table 6. Lateralized EEG characteristics in schizophrenia: effects of motor and cognitive tasks

---

Peculiar hemispheric modifications during specific cognitive tasks (1,11)

Peculiar right hemisphere modifications during spatial tasks (28)

Peculiar left hemisphere modifications during motor tasks (23)

No peculiar hemisphere modifications during motor tasks (25)

129

Some studies have shown that the lateralized malfunctioning of patients receiving drug therapy is not substantially different from that of drug-free patients (3) On the other hand, a direct influence of neuroleptic treatment on hemispheric functional asymmetries has been reported (25) Our own experience indicates that the drug regime is crucial for explaining EEG power characteristics. In a recent study (29), Multiple Linear Regression Analysis was used in order to evaluate the effects of some clinical characteristics (independent variables) such as diagnostic subtype, symptomatology and drug treatment, on the variance of the log-transformed values of the EEG power of each frequency band (i.e., delta 1-4, theta1 4-6, theta2 6-8, alpha1 8-10, alpha2 10-13, beta1 13-18, beta2 18-30). The variance of the energy values under eyes-closed conditions was significantly explained for delta, alpha1 and alpha2 bands primarly by the variable drug treatment.

Conclusions

The tentative conclusions to be drawn from the results presented are as follows:

a) As compared to normal subjects, the physiological aspects of cerebral asymmetries seem to be altered in most schizophrenic patients.

b) Hemispheric dysfunction might be related to some abnormalities in a specific cerebral area or to some disturbances at a more general and non- specific intregrative level.

c) Some aspects of lateralized deficits in schizophrenia are probably due to the genetic and inherent aspects of the disease.

d) Sex, age at onset and duration of the illness and psychopathological characteristics should be taken into account in the evaluation of research results.

e) Psychopharmacological treatment seems to be another important variable in determining cerebral functional disorders.

f) Variability in performance ability as well as motivational level and cooperativeness are factors that should not be neglected when interpreting findings.

An unresolved question is whether cerebral assessment reflects some specific hemispheric malfunctioning, a modification of the highest levels of cerebral organization, or both.
Neurophysiological as well as neuropsychological data indicate the differing involvement of the right and left hemispheres as this relates to duration of illness and psychopathological symptomatology. This leads to the further problem of whether the modification of lateralized specializations in schizophrenia is a trait characteristic of the disease. On the other hand, some of the neuropsychological and neurophysiological aspects of hemispheric specialization seem to be more highly influenced by the patients' state and symptoms. If this is the case, the influence of arousal or sensory processing disorders in schizophrenic hemisphere dysfunction has to be carefully investigated. Perinent to this issue are recent findings suggesting defective interhemispheric transfer of informations in schizophrenia (30). Although anomalies in cerebral hemispheric asymmetries are probably a normal variant in psychiatric patients, they could be representative of neurointegrative deficits (31). The general problem of the difficulty of sampling homogeneous groups of patients, high inter and intra-subject variability, as well as assessment of the subjects level of motivation and his cooperativeness are all factors undoubtedly complicating the interpretation of results.

References
1. Flor-Henry, P (1979). Commentary on the cortical issues and neuropsychological and elettroencephalografic findings. In: Gruzelier, J and Flor-Henry, P (eds) "Hemispheric asymmetries of function in psychopathology". p. 639. (Amsterdam: Elsevier-/North-Holland Biomedical Press)
2. Coger, RW, Dymond AM, Serafetinides EA (1979). Electroencephalographic similarities between chronic alchoolics and chronic, nonparanoid schizophrenics. Arch Gen Psychiatry, 36, 91
3. Morihisa JM, Duffy, FH, Wyatt, RJ (1983). Brain electrical activity mapping (BEAM) in schizophrenic patients. Arch Gen Psychiatry, 40, 719
4. Mukundan, CR (1986). Computed EEG in schizophrenics. Biol Psychiatry, 21, 12225
5. Luchins, DJ, Weinberger, DR, Wyatt, RJ (1979). Schizophrenia: Evidence of a subgroup with reversed cerebral

asymmetry. Arch Gen Psychiatry, 36, 1309

6.   Abrams, R, Redfield, J, and Taylor, MA (1981). Cognitive dysfunction in schizophrenia, Affective Disorders and Organic Brain Disease. Brit J Psychiatry, 139, 190

7.   Gur, RE (1978) Left hemisphere dysfunction and left hemisphere overactivation in schizophrenia.   J Abn Psychology, 87, 226

8.   Gruzelier, JH (1981).   Cerebral laterality and psychopathology: facts and fiction. Psychol Medicine, 11, 219

9.   Etevenon, P, Pidoux, P, Rioux, P, Peron-Magnan, P, Verdeaux, G and Deniker, P (1979). Intra and Interhemispheric EEG differences quantified by spectral analysis. Comparative study of two groups of schizophrenics and a control group. Acta Psychiat Scand, 60, 57

10.   Stevens, JR, and Livermore, A (1982). Telemetered EEG in schizophrenia: spectral analysis during abnormal behaviour episodes. J Neurol Neurosurg Psychiatry, 45, 385

11.   Shagass C, Roemer RA and Straumanis JJ (1982). Relationships between psychiatric diagnosis and some quantitative EEG variables. Arch Gen Psychiatry, 39, 1423

12.   Venables, P and Fleminger, JJ (1980). Primary dysfunction and cortical lateralization in schizophrenia.   In: Koukku, M, Lehmann, D, and Angst, J (eds) "Functional states of the brain: their determinants". p.243 (Elsevier/North-Holland Biomedical Press)

13.   Gambini, O., Cazzullo CL and Scarone , S (1986). Interpretation of abnormal responses to the Quality Extinction Test in schizophrenia. J Neurol Neurosurg Psychiatry, 49, 997

14.   Schwartz, AS, Marchok, PL, Kreinick, CJ and Flynn, RE (1979). The asymmetric lateralization of tactile extinction in patients with unilateral cerebral dysfunction. Brain, 102, 669

15.   Scarone, S, Pieri, E, Gambini, 0, Massironi, R and Cazzullo, CL (1982). Brit J Psychiat 141, 350

16.   Scarone, S Gambini, O and Pieri, E (1983). Dominant hemisphere dysfunction in chronic schizophrenia: Schwartz test and short aphasia screening test. In: Flor-Henry P and Gruzelier JH (eds) "Laterality and Psychopatology". p. 129-142 (Elsevier Science Publishers B.V.)

17.   Scarone, S, Cazzullo, CL and Gambini, 0 (1987). Asymmetry of lateralized hemispheric functions in schizophrenia.   Br J Psychiatry,p in press

18.   Klass DM and Daly DD (1979). Current Practice of Clinical Electroencephalography. Raven Press

19.   Cooley JW and Tukey JW (1965). An algorithm for the machine calculation of complex fourier series. Math Comput, 19, 297

20.   Itil TM, Saletu B and Davis S (1972). EEG findings in chronic schizophrenics based on digital computer period analysis and analog power spectra. Biol Psychiatry, 5 ,1

21.   Flor-Henry, P, Koles RJ, Horwarth BG and Burton, L (1979). Neurophysiological   studies   of   schizophrenia,   mania   and depression. In Flor-Henry P and Gruzelier J (eds) "Laterality and Psychopathology". p 189-222 (Elsevier Science Publishers B.V.)

22.   Etevenon P, Peron-Magnan P, Rioux P , Pidoux, B, Bisserbe, JC, Verdeaux, G and Deniker, P (1981). Schizophrenia assessed by computerized EEG. Adv Biol Psychiatry, 6,29

23.   Guenter, W and Breitling D (1985). Predominant sensorimo-tor area left hemisphere dysfunction in shizophrenia measured by brain electrical activity mapping. Biol Psychiatry 20, 515

24.   Venables PH   (1980) Primary dysfunction and cortical lateralization in schizophrenia. In Koukkou M, Lehmann D and Angst J (eds) "Functional states of the brain: their determi-nants". p 243-264 (Elsevier/North Holland Biomedical Press)

25.   Merrin, EL, Fein G, Floyd TC and Yingling CD (1986). EEG asymmetry in schizophrenic patients before and during neuroleptic treatment. Biol Psychiatry, 21, 455

26.   Etevenon, P, Pidoux P, Rioux, P and Deniker, P. (1979) Intra and inter-hemispheric EEG differences quantified by spectral analysis: Comparative study of two groups of schizophrenics and a control group. Acta Psychatr Scand, 60, 57

27.   Serafetinides, EA, Coger, RW and Martin, J (1981). Schizophrenic symptomatology and cerebral dominance patterns: A comparison of EEG, AER and BPRS measures. Compr Psychiatry, 22, 218

28.   Morstyn, R, Duffy, FH and McCarley RC (1983). Altered topography of EEG spectral content in schizophenia. Electroencep-halog Clin Neurophysiol, 56, 263

29.   Scarone, S, Pugnetti, L;, Cattaneo, AM, Biserni, P., Capuano, P., Colombo, C, Cattaneo, R., Gambini, O. and Cazzullo, CL (1986). Electrophysiological indexes of hemispheric abnormalities in schizophrenia: clinical and epidemiological correlates. In Shagass C et Al (eds) "Biological Psychiatry 1985". p 1054-1056 (Elsevier Science Publishing Co)

30.   Carr, SA (1980) Interhemispheric transfer of stereognosic information in chronic schizophrenia. Brit J Psychiat, 136, 53
31.   Baribeau-Braun, J, Picton, TW and Gosselin JY (1983). Schizophrenia: A neurophysiological evaluation of abnormal information processing. Science, 219,874

14
In vivo studies of dopamine receptors in schizophrenia

G. Sedvall

INTRODUCTION

The discovery of the therapeutic effect of neuroleptic drugs in patients with schizophrenic psychosis and the subsequent demonstration
that neuroleptic drugs have profound effects on dopamine metabolism
in the brain early focussed the interest to a possible role of
dopaminergic mechanisms in the pathophysiology of schizophrenia (1,
2). Since the neuroleptic drugs later have been shown to induce
their effect on dopamine metabolism in the brain by blocking the
receptors for this transmitter, it has generally been assumed that
there may be an overstimulation of dopaminergic transmission in some
schizophrenic patients (2). Support for this view was supplied by
some studies demonstrating increased concentrations of the major
dopamine metabolite, homovanillic acid, in the cerebrospinal fluid
and plasma of selective subgroups of schizophrenic patients (3,4).
However, many investigators failed to reproduce these findings. A
more robust biochemical finding in schizophrenia was the demonstration that the major dopamine containing areas in the brain of
diseased schizophrenic patients contain an abnormally elevated
number of dopamine receptors as compared to brains from non schizophrenic controls (5). These findings have stimulated a number of
investigators to initiate a detailed mapping of possible biochemical
abnormalities in the schizophrenic brain obtained postmortally.
Since deviations of biochemical functions in post mortem tissue may
be related to postmortal alterations and since many receptor systems
may be altered following the long term in vivo treatment of schizophrenic patients with neuroleptic drugs it seems highly desirable to
complement postmortem studies by direct in vivo observations in
living patients.
    Positron emission tomography is a recently developed technique
which has the potential to examine some biochemical functions in the
living human brain. Thus, recently methods have been developed for
the quantification of glucose and oxygen metabolism in the brain by
positron emission tomography (6). We have recently focussed our
interest on the possibility to study neuroreceptor characteristics
in the living human brain using the technique of positron emission
tomography (7-9). Wagner et al 1983 in pioneering experiments
was first to demonstrate the possibility to visualize dopamine

receptors in the brain by positron emission tomography (10). This task was accomplished by using a ligand that binds to dopamine receptors and to demonstrate the accumulation of this compound labelled with the positron emitting isotope carbon-11 (11-C) in dopamine receptor rich brain areas by positron emission tomography. The ligand Wagner et al used was N-methylspiperone which is a rather non selective ligand with regard to dopamine receptor binding. We have been interested in developing specific ligands binding to the different subtypes of dopamine receptors that recently have been shown to occur. There are at least two types of dopamine receptors in the brain, the dopamine D1 and the D2 receptors. We have recently been successful in labelling specific ligands for these two receptor types with C-11 and demonstrated their usefulness for studying the distribution and biochemical characteristics of the dopamine D1 and the dopamine D2 receptors in the living brain. These techniques will be applied to the analysis of dopamine receptor characteristics in the brain of schizophrenic patients.

METHODS

^{11}C-SCH 23390 and ^{11}C-raclopride, selective ligands for dopamine D1 and D2 receptors respectively were prepared using the desmethylated analogues of the ligands as precursors for methylation using 11-C-labelled methyl iodide (11,12). A local cyclotron placed at our hospital was used for production of 11-C. The compounds were injected intravenously into healthy volunteers and drug naive and drug treated schizophrenic patients. Local radioactivity in brain regions and plasma was followed for about one hour. The positron camera system PC 384 at the Dept of Neuroradiology, Karolinska hospital, was used (13).

RESULTS

In healthy volunteers both ligands accumulated markedly in the dopamine receptor rich basal ganglia (14). In these regions the radioactivity was three to fourfold higher than in other brain regions. After both ligands the cerebellum exhibited a low level of radioactivity. The cerebellum is known to contain a negligible concentration of dopamine receptors. After 11-C-SCH 23390 but not 11-C-raclopride there was also a conspicuous accumulation of radioactivity in the neocortex but not in the cerebellar cortex. These results indicate that it is possible to visualize dopamine D1 and dopamine-D2 receptor binding in the brain using these ligands and the technique of positron emission tomography. The results indicate that also in vivo there is marked dopamine receptor binding in the basal ganglia. There also seems to be receptor binding in the neocortex after administration of the D1 specific ligand 11-C-SCH 23390.

In an attempt to quantify dopamine receptor characteristics a saturation analysis in vivo was performed using 11-C-raclopride as the ligand (13). In this study increasing doses of non labelled raclopride was administered together with a constant amount of 11-C-raclopride in a series of experiments. It was shown that the specific labelling in the basal ganglia using cerebellum as a region for non specific binding, showed saturation kinetics. From the binding isoterms Bmax and Kd-values could be calculated in healthy volunteers and schizophrenic patients. The Bmax values were in the order of 20 pmol/g putamen tissue. The Kd-values were in the order of 4 nM (13).

In schizophrenic patients treated with conventional doses of chemically different types of antipsychotic drugs there was a marked diminution of 11-C-raclopride binding in the basal ganglia (13,14). Thus, after treatment of conventional neuroleptic drugs as chlorpromazine, haloperidol, perphenazine, and thioridazine there was a more than 70% reduction of specific binding in the basal ganglia. Interestingly, enough also after non conventional antipsychotic drugs as clozapine, sulpiride and melperone there was a more than 70% reduction of 11-C-raclopride binding (Farde, Wiesel and Sedvall, unpublished data). On the other hand there was no or only a slight diminution of radioactivity in the basal ganglia after administration of 11-C-SCH 23390 in the drug treated patients (15). Flupenthixol, a neuroleptic, known to have D1 receptor blocking properties, show only an about 10% reduction of 11-C-SCH 23390 binding in the brain in a patient treated with conventional doses of this drug. These results indicate that all the major classes of antipsychotic drugs induce a considerable blockade of dopamine D2 receptors in the brain during treatment with conventional doses. None of the hitherto available antipsychotic drugs seem to have any marked effect on dopamine D1 receptor binding in the brain.

So far about 10 drug naive schizophrenic patients have been examined using 11-C-raclopride as the ligand. A comparison to eight age-matched healthy volunteers indicates that there is an elevation of dopamine D2 receptor binding in the brain of the schizophrenic patients. However, there are major individual differences. Since the evaluation of these results has not been completed we expect to analyze another ten patients before any definite conclusions can be drawn with regard to the clinical importance of the alterations of dopamine receptor characteristics in schizophrenic patients.

COMMENTS

The results summarized in this communication demonstrate the potential of using specific ligands for studying dopamine receptor characteristics in the living human brain using the technique of positron emission tomography. Such in vivo studies should be an excellent complement to the previously available postmortem studies on such receptors. Since the in vivo approach examines the receptors in a true physiological environment such studies may possibly demonstrate differences in comparison to the in vitro approach.

The clinical importance of this approach has already been indicated by the demonstration of the pronounced dopamine D2 receptor occupancy in the schizophrenic patients treated with conventional doses of antipsychotic drugs. Before these studies se studies were performed the extent to which dopamine receptors are blocked during clinical treatment was unknown.

The possibility to quantify the receptors using the saturation analysis also gives a direct possibility to study quantitatively receptor characteristics not only in patients with schizophrenia but also in other categories of neuropsychiatric disorders.

The major limition of the positron camera technique is the resolution. The instrument by which the present studies were obtained has a resolution in the order of about 7 mm. The next generation of cameras will be expected to have a resolution of 4 mm. This will give the possibility to study dopamine receptor characteristics also in other brain areas besides the major basal ganglia. Thus, structures as the limbic cortex, neocortex, the amygdala and the hippocampus will be major targets for research in the area of dopamine receptors.

The preliminary indication of an elevated number of dopamine D2 receptors in the brains of drug naive schizophrenic patients seems promising and confirms the previous in vitro studies. The pathophysiological relevance of this alteration has now to be explored by correlating the dopamine recetpor data to a number of clinical data.

During the years to come there are reasons to believe that an increasing number of useful ligands for PET visualization of several other neuroreceptor systems in the human brain besides dopamine receptors will be developed. This new methodology may ultimately turn out to be a promising new approach to biochemical diagnostic systems in psychiatry.

ACKNOWLEDGEMENTS

I gratefully acknowledge the skilful typing assistence of Mrs Birgit Lönn. Grants from The Swedish Medical Research Council (B87-21X-03560-16C), the Bank of Sweden Tercentenary Foundation, the National Institute of Mental Health (MH 41205-1) and the Karolinska Institute supported the studies summarized in this paper.

REFERENCES

1.   Carlsson, A (1978) Antipsychotic drugs, neurotransmitter receptors, and schizophrenia. Am J Psychiatry, 135, 164-173
2.   Snyder, SH (1984) Drug and neurotransmitter receptors in the brain. Science, 224, 22-31
3.   Sedvall, G and Wode-Helgodt B (1980) Aberrant monoamine metabolite levels in CSF and family history of schizophrenia. Arch Gen Psychiatry, 37, 1113-1116
4.   Wiesel, F-A, Bjerkenstedt, L, Herlofsson, C, Härnryd, C, Nybäck, H, Oxenstierna, G and Sedvall, G (1984) Psychiatric morbidity within the family and steady-state levels of cerebrospinal fluid monoamine metabolites. In: Usdin, E, Dahlström, A, Carlsson, A, Engel, J eds. Catecholamines: Neuropharmacology and Central Nervous System - Therapeutic Aspects, p 131-138, Alan R, Liss Inc. 5.
Seeman, P, Ulpian, C, Bergeron, C, Riederer, P, Jellinger, K, Gabriel, E, Reynolds, GP, Tourtellotte, WW (1984) Bimodal distribution of dopamine receptor densities in brains of schizophrenics. Science, 225, 728-731
6.   Phelps, ME and Maziotta, JC (185) Positron emission tomography: Human brain function and biochemistry. Science, 228, 799-809
7.   Sedvall, G, Blomqvist, G, dePaulis, T., Ehrin, E., Eriksson, L, Farde, L, Greitz, T, Hedström, C-G, Ingvar, DH, Litton, J-E,

L, Farde, L, Greitz, T, Hedström, C-G, Ingvar, DH, Litton, J-E, Nilsson, JLG, Stone-Elander, S, Widén, L, Wiesel, F-A and Wik, G (1983) 11-C labelled benzamindes as ligands for imaging of dopamine receptors in the human brain. ACNP Annual Meeting, December. San Juan, 1983

8.   Farde, L, Ehrin, E, Eriksson, L, Greitz, T, Hall, H, Hedström, Litton, J-E and Sedvall, G (1985) Substituted benzamides as ligands for visualization of dopamine receptor binding in the human brain by positron emission tomography. Proc Natl Acad Sci USA, 82, 3863-3867

9.   Sedvall, G, Farde, L, Stone-Elander, S and Halldin, C (1986) Dopamine D1-receptor binding in the living human brain. In: Breese, GR and Creese, I eds.) "Adv Exp Med Biol: Neurobiology of central D1-dopamine receptors". Vol 204, p 119

10.   Wagner, HN, Bruns, HD, Dannals, RF, Wong, DR, Långström, B, Duelfer, T, Frost, JJ, Ravert, HT, Links, JM, Rosenbloom, SB, Lukas, SE, Kramer, AV and Kuhar, MJ (1983) Imaging dopamine receptors in the human brain by positron tomography. Science, 221, 1264-1266 11.

11.   Ehrin, E, Farde, L, DePaulis, T, Eriksson, L, Greitz, T, Johnström, P, Litton, J-E, Nilsson, JLG, Sedvall, G, Stone-Elander, S, Ögren, S-O (1985) Preparation of 11-C-labelled raclopride, a new potent dopamine receptor antagonist: Preliminary PET studies of cerebral dopamine receptors in the monkey. Int J Appl Radiat Isot 36, 269-273

12.   Halldin, C, Stone-Elander, S, Farde, L, Ehrin, E, Fasth, K-J, Långström, B and Sedvall, G (1986) Preparation of 11-C-labelled SCH 23390 for the in vivo study of dopamine D1 receptors using positron emission tomography. Int J Appl Radat Isot, 37, 1037-1043

13.   Farde, L, Hall, H, Ehrin, E and Sedvall, G (1986) Quantitative analysis of D2 dopamine receptor binding in the living human brain by PET. Science, 231, 258-261

14.   Sedvall, G, Farde, L, Persson, A and Wiesel, F-A (1986) Imaging of neurotransmitter receptors in the living human brain. Arch Gen Psychiatry, 43, 995-1005

15.   Farde, L, Halldin, C, Stone-Elander, S and Sedvall, G (1987) PET analysis of human dopamine receptor subtypes using 11-C-SCH 23390 and 11-C-raclopride. Psychopharmacology, In press.

15
Nuclear magnetic resonance imaging in schizophrenia: a preliminary study

A. Rossi, P. Stratta, M. Casacchia,
M. Gallucci and R. Passariello

INTRODUCTION

The study of brain morphology in recent years has primarily used computed tomography (CT) scan and anatomical or histological examinations of brains of deceased patients. CT studies identified dilatation of ventricles of the brain and others structural abnormalities in psychiatric disorders, especially in patients with schizophrenia |1|. Among others the choice of control group represent a great source of variability in the results of studies of Lateral Ventricular Enlargement (LVE) in schizophrenia |2|. Technical limitations of CT studies are represented by ionizing radiation, bone artifact and low soft tissue contrast. These problems may be overcome with the safer Magnetic Resonance Imaging. In addiction MRI produces images of greater anatomical detail than CT scans, and allows imaging of the brain in multiple planes.

Relatively few studies have addressed these issues |3, 4, 5, 6|.

We undertook a MRI study in schizophrenia and report here results of area and image intensity measurements of several brain structures in a group of young schizophrenic patients and normal subjects.

METHOD

7 chronic schizophrenic patients, 2 females and 5 males, (according with DSM III criteria), were studied by means of clinical and MRI examinations. The mean age was 34.71 years (SD 4.92).

Midsagittal and axial MRI scans were also obtained on 7 healthy volunteers. Normal controls were chosen from among employees and relatives of the hospital staff and were individually matched for age (within 3 years) and sex with the schizophrenic patients. The mean age was 32.71 years (SD 6.42).

Subjects were excluded if they manifested recent evidence or had history of alcohol or drug abuse, or evidence of neuro-logical disease, morphological insult to the brain, or had surgical metal or electronic implants considered able to inter-fere with MRI evaluation.

MRI examinations were performed by means of an Ansaldo Esatom 5000 scanner operating at a 0.5 Tesla magnetic field. "T1" weighted (SE 350/30 on sagittal plane and 1800/30 on the axial one) and "T2" weighted (SE 1800/120) sequences were obtained on sagittal and trasversal planes (10 mm slice thickness) at 15° to the orbitomeatal line. All linear and area measurements were made by a neuroradiologist (M.G.) who was unaware of the diagnosis.

The mid-sagittal slice which gave the clearest outline of the corpus callosum (CC) was used for measurement purpose.

The total surface of the cortical area, corpus callosum and fourth ventricle were measured in square centimeters. The anterior-posterior lenght of the corpus callosum was also performed.

Image intensity was measured from both SE 1800/30 and SE 1800/120 scans. Image intensity was measured in speci-fic areas (see Tab.1) with a region of interest (ROI) of 43 pixels. The analyses reported in this paper were performed using "standardized" imaging intensities. A standard (provided by Ansaldo-Esacontrol) was measured on the same scan.

Standardized image intensity was calculated as the ratio of the image intensity of the brain area sampled to the image intensity of standard.

VBR (Ventricular Brain Ratio) was calculated at the assial scan as ratio of the area of lateral ventricles at their largest to that of the entire brain in the section, expressed as a percentage. The maximum width between right and left anterior horns, between right and left occipital horns and the intercaudatus diameter were also performed. All the measu-rements were made from the computer console.

Pearson product-moment was performed for correlation analysis. Two-tailed t-test was used for comparisons.

RESULTS

Schizophrenics showed a statistically significant larger VBR (p< 0.05) than controls. Interestingly normal controls showed a significantly higher corpus callosum brain ratio and a larger corpus callosum area (p<0.05) (Tab.1).

Table 1. Linear and area measurement in schizophrenics and controls

	Schizophrenics (N° = 7)		Controls (N° = 7)		p
	Mean	SD	Mean	SD	
A:	6.56	2.45	4.26	1.24	0.05
B:	6.1	1.21	7.6	1.24	0.05
C:	7.43	1.06	9.14	1.32	0.05
D:	7.59	0.71	7.58	0.28	NS
E:	81.54	5.03	83.32	8.69	NS
F:	0.94	0.47	0.71	0.38	NS
G:	3.21	0.44	3.53	0.39	NS
H:	5.8	0.65	5.59	0.64	NS
I:	1.32	0.38	1.28	0.26	NS

A – VBR
B – Corpus Callosum area (cm2)
C – Corpus Callosum brain ratio
D – Corpus Callosum anterioposterior lenght (cm)
E – Cortical Sagittal Area (cm2)
F – 4th Ventricle Area (cm2)
G – Maximum width between right and left anterior horns (cm)
H – Maximum width between right and left occipital horns (cm)
I – Intercaudatus diameter (cm)

Although there were no statistically significant differences in image intensity between schizophrenics and controls, schizophrenics showed lower image intensity in both gray and white matter in each ROI measured (Tab.2).

Age was highly correlated with corpus callosum area (r=0.682) and with VBR (r=0.58) in schizophrenic group but did not reach the statistical significance; age did not show any valuable relationships with corpus callosum area and VBR (r=-0.179 and r=0.22 respectively) in normal controls.

Table 2. Differences between schizophrenics **and** **normal** controls for standardized SE–30 and SE–120 axial images

SE–1800/30 Axial	Schizophrenics (N°7)		Controls (N°7)			
	Mean	SD	Mean	SD	t	p
Prefrontal white matter						
Right	.57	.08	.66	.31	.70	NS
Left	.61	.08	.67	.29	.49	NS
Prefrontal gray matter						
Right	.64	.09	.73	.35	.64	NS
Left	.64	.08	.74	.31	.77	NS
Anterior Temporal white matter						
Right	.68	.07	.80	.40	.78	NS
Left	.72	.08	.85	.41	.79	NS
Anterior Temporal gray matter						
Right	.70	.09	.88	.44	1.04	NS
Left	.75	.07	.89	.40	.91	NS
SE–1800/120 Axial						
Prefrontal white matter						
Right	.92	.40	1.49	1.07	1.30	NS
Left	.95	.39	1.71	1.38	1.40	NS
Prefrontal gray matter						
Right	1.54	.66	2.36	1.71	1.17	NS
Left	1.37	.59	2.32	2.00	1.20	NS
Anterior Temporal white matter						
Right	1.47	.62	2.13	1.82	.90	NS
Left	1.46	.59	2.34	1.81	1.21	NS
Anterior Temporal gray matter						
Right	1.75	.76	2.94	2.40	1.25	NS
Left	1.89	.79	3.01	2.31	1.20	NS

DISCUSSION

MRI scans of the brain of schizophrenics patients indicated that they can be imaged with greater clarity and detail by MRI. Nasrallah et al. |4, 5|, Olson et al. |7| reported in studies of linear and area measurements that the CC in schizophrenic patients is different than controls and that there are gender related differences in the morphology of the ventricular system. Nasrallah et al. |5| and Mathew et al. |6| reported an increased callosal thickness and longer corpora callosa in schizophrenia, respectively.

Our data do not confirm these results and on the opposite we report a larger CC area and a larger CC brain ratio in normals; nevertheless our mid-sagittal results are very close to those reported by Nasrallah et al. |4|.

Smith et al. |8|, using a biological parameter, image intensity, reported differences of the brain white and gray matter in selected brain areas in schizophrenics versus controls in the IR-30 scanning but not in SE-30 mode.

Our study confirm the conclusion of Smith et al. |8| that the SE scanning mode do not detect statistically significant differences between groups but the observation that schizophrenics show lower image intensity measures in each ROI deserves further investigations.

The use of imaging of the brain in multiple planes and the study of image intensity in the same clinical sample may add further clues to knowledge of the cerebral abnormalities reported by NMR studies. From these preliminary results we are moving in this direction.

REFERENCES

1. Johnstone, EC, Crow, TJ, Frith, CD, et al. (197 6). Cerebral ventricular size and cognitive impairment in chronic schizophrenia. Lancet, 2, 924
2. Smith, GN and Iacono, WG (1986). Lateral ventricular size in schizophrenia and choice of control group.Lancet, 21, 1450
3. Andreasen, N, Nasrallah, HA, Dunn, V, et al. (1986). Structural abnormalities in the frontal system in schizophrenia. A Magnetic Resonance Imaging study. Arch Gen Psychiatry, 43, 136
4. Nasrallah, HA, Andreasen, NC, Coffman, JA, et al. (1986). Morphology of the corpus callosum in schizophrenia.

Fifth Annual Meeting of the Society of Magnetic Resonance in Medicine. August 19–22, Montreal, Quebec Canada. Book of Abstracts p 705

5. Nasrallah, HA, Andreasen, NC, Coffman, JA, et al. (1986a). A controlled Magnetic Resonance Imaging Study of Corpus Callosum Thickness in Schizophrenia. Biol Psychiatry, 21, 274

6. Mathew, RJ, Partain, CL, Prakash, R, et al. (1985). A study of the septum pellucidum and corpus callosum in schizophrenia with MR imaging. Acta Psychiatr Scand, 72, 414

7. Olson, SC, Nasrallah, HA, Andreasen, NC, et al. (1986). Brain morphology in schizophrenia by magnetic resonance imaging. Fifth Annual Meeting of the Society of Magnetic Resonance in Medicine. August 19–22, Montreal, Quebec Canada. Book of Abstracts p 703

8. Smith, RC, Baumgartner, R, Calderon, M, et al. (1985). Magnetic Resonance Imaging studies of Schizophrenia. Psychopharmacology Bulletin, 21, 3, 558

BIOCHEMISTRY

16
Plasma opioid levels: a possible marker of neurotransmitter impairments in schizophrenia

F. Brambilla, F. Facchinetti,
F. Petraglia and A.R. Genazzani

The data on the periferal secretion of opioid peptides in schizophrenia are contradictory.Normal,increased or decreased plasma
levels of opioid-like material or beta-endorphin(β-EP) have been
reported(1-10). The conflicting results may be due to methodological
factors,including differences in patient selection , the type
of assay used to measure the peptides, the possible interference
of previous pharmacological treatments. The meaning of the abnormalities observed is even more controversial. It has been suggested
that an altered peripheral opioid secretion may be the expression
of a generalized phenomenon representing one of the possible causes
of the mental disorder .Data on the opioid-like material levels in
the cerebrospinal fluid(CSF),which should better reflect brain secretion of the substance,are also contradictory Again,low,normal
or high levels have been reported,with CSF and plasma values sometimes concordant and sometimes not(1-4,9,11-26). It has been
suggested that CSF and plasma levels of opioids may be discordant
because blood values mainly reflect the pituitary and CSF the brain
secretion,the two pools being possibly subjected to different regulatory mechanisms and having different methabolic pathways.In this
line,alterations of the secretory pattern of peripheral opioids
should not be the cause of or interfere with the development of
schizophrenia. However, the peripheral opioid secretion may influence the brain function in spite of the low permeability of the blood-
brain barrier(BBB) to these peptides.Peripheral opioids could still
reach the brain centers through the portal vessels by reverse blood
flow,through areas at which the BBB is weakly formed or through
cleavage of the peptides to smaller fragments which cross the BBB
while still retaining activity at the neuronal levels(27-29).
Consequently, peripheral opioids might influence the development,
course and prognosis of schizophrenia. At the moment,however,the
two compartments,central and peripheral,should be considered as

two separate entities with regard to opioid secretion,each one with
its specific function.

Peripheral secretion may reflect alterations of the brain regulatory neurotransmitter systems and therefore mirror the biochemical alterations at the basis of the psychosis,as has been suggested
for many other hormonal parameters.In this context,we studied the
basal and dynamic secretory patterns of β-EP and beta-lipotropin
(β-LPH) as possible markers of brain neurotransmitter function in
a group of chronic schizophrenic patients.

MATERIAL AND METHODS

Forty-five chronic schizophrenics,31 hebephrenics and 14 paranoids,
diagnosed according to the DSM III,were investigated.Twenty-seven
were men and 18 women (13 fertiles and 5 postmenopausal),aged 23
to 64 years,with 5 to 31 year histories of the disease.All the
subjects were inpatients at our Institute.They all had had various
neuroleptic,benzodiazepine and electroschock terapies,but had not
received any type of treatment for at least 10 days prior to the
study.

Baseline plasma β-EP and β-LPH levels were assayed for all
the patients in blood collected at 9.00 a.m.Intravenous insulin
stimulation tests(0.1 I.U./kg bw)were given to 8 patients,with β-EP
and β-LPH levels measured at 15-30 min intervals for 2 hours after
the stimulation.The same 8 patients had the two peptides measured
at 9.00 a.m. and at 6.00 p.m. so as to evaluate their circadian
pattern of secretion.The same 8 patients received a multidose dexamethasone suppression test(0.5 mg every 6 hrs for 48 hrs) and the
two peptides were measured at 9.00 a.m. before the administration
of dexamethasone and 2 days later at 9.00 a.m.,8 hrs after the last
dose.

In 8 patients β-EP and β-LPH were measured after 10 days of Haloperidol therapy(1 mg/kg bw per os),in other 8 after 30 days of Haloperidol therapy(10-18 mg/day per os) and in other 9 after 14-18
months of a treatment with Haloperidol(10-30 mg/day per os),combined
with either chlorpromazine(25-75 mg/day per os),or clotiapine (60
mg/day per os),or fluphenazine decanoate (25-50 mg /month i.m.).

Thirty-five healthy volunteers,defined as psychologically normal after psychiatric examination and matched for sex and age acted
as controls.Informed consent was obtained from both patients and
controls.

β-EP and β-LPH plasma levels were measured by RIA,after extrac-
tion from plasma through a G-75 Sephadex column to separate the
two peptides according to the method of Facchinetti et al.(30).

The data were analyzed statistically by the Student T test for
paired and unpaired data.

RESULTS

Basal β-LPH values were elevated in 23 patients(9-22 fmol/ml) and
normal in the other 22 patients.Mean values for the entire patient
group were significantly higher than for the controls(M+ SE= 10.5
± 0.8 versus 5.9 ± 0.3 fmol/ml ; P < 0.01). Basal β-EP levels were
elevated in 30 patients (9-72 fmol/ml) and normal in the other 15.
The mean values for the schizophrenics were significantly higher
than for the controls (M+ SE 18.0 ± 2.2 versus 6.2 ± 0.3 fmol/ml;
P < 0.01)(Fig.1.)

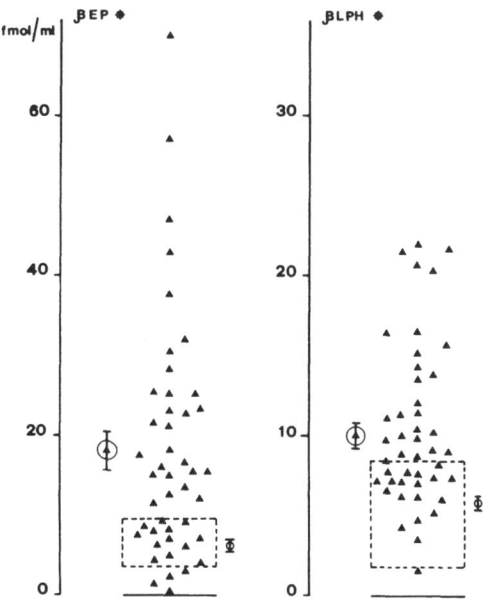

FIGURE 1   Baseline β-EP(left pannel) and β-LPH(right pannel) plasma
    levels   of 45 chronic schizophrenics (▲ ; M+ SE ⏀)and of 35
    controls(M+ SD ⌐ ‾ ‾ ‾ ‾ ¬ ;M+ SE ⏀). * P < 0.01

151

Fifteen of the 45 patients had concomitantly high levels of β-EP
and β-LPH ,and 9 concomitantly normal levels,while in the others
there was no correlation between the levels of the 2 peptides.There
was no correlation between β-EP and β-LPH levels and type of schizophrenia,age or sex of the patients,duration of the disease or
type and duration of previous pharmacological therapies.

Insulin did not stimulate β-EP and β-LPH plasma levels in 3
patients but the other 5 responded normally(Fig.2).

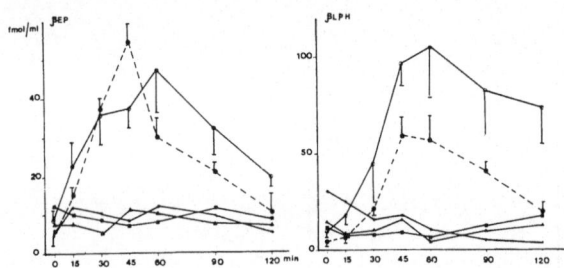

FIGURE 2   β-EP(left pannel) and β-LPH (right pannel) plasma levels
    after insulin administration to 8 chronic schizophrenics
    (M ± SE of 5 patients o——o ,individual data of 3 patients
          ●——●) and to 8 controls(M ± SE ●-----●)

Circadian rhythmicity of β-EP was impaired in 2 cases ,that of β-LPH
in 1 case, and normal in all the other patients(Fig.3).
Response to dexamethasone inhibition was blunted for β-EP in 5 cases
and for β-LPH in 6 (Fig.4).

The administration of Haloperidol for either 10 or 30 days and
of the mixed neuroleptic therapy for 14-18 months did not significantly modify the mean values of either β-EP and β-LPH.Even though
increased or decreased levels of the two peptides were observed

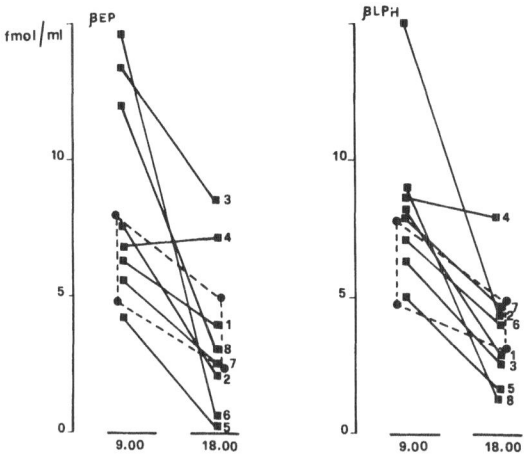

FIGURE 3   Individual plasma levels of β-EP(left pannel)and β-LPH
(right pannel) of 8 chronic schizophrenics(•——•) and M +
SD   values of 8 controls (⌐----⌐) at 9.00 a.m. and 6.00
p.m.

in single patients at the end of the treatment when compared to
their levels before treatment,mean values did not change (Haloperidol for 10 days= β-EP = 32.1 + 8.2 versus 27.0 + 9.6 fmol/ml;β-LPH=
16.4 + 2 versus 16.6 + 1.3 fmol/ml .Haloperidol for 30 days= β-EP=
20.1 + 5.2   versus 25.4 + 3.1 fmol/ml; β-LPH= 8.5 + 1.8 versus
12.2 + 2.9 fmol/ml. Mixed neuroleptic therapy= β-EP = 13.8 + 1.3
versus 13.3 + 1.3 fmol/ml). A significant increase of β-LPH levels
(M+ SE = 8.0 + 0.6 versus 15.2 + 2.4 fmol/ml ; P ＜ 0.01) was
observed only in the group of patients who received the mixed
neuroleptic therapy for 14-18 months.No correlations were observed
between changes of the levels of the two peptides in single patients
and the clinical effects of the therapies

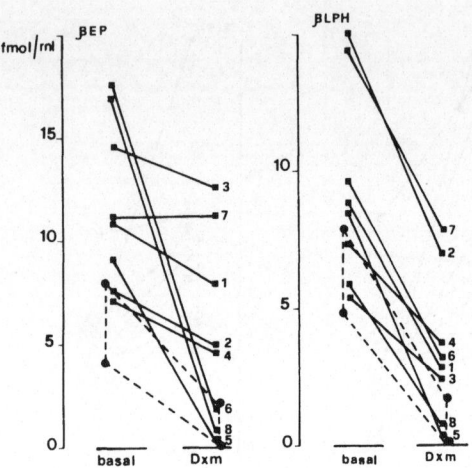

FIGURE 4 Individual plasma levels of β-EP(left pannel) and β-LPH
(right pannel) of 8 chronic schizophrenics before and after
dexamethasone administration (•————•) and M ± SD values
of controls (┌─────┐)

DISCUSSION AND CONCLUSIONS

The present data indicate that β-EP and β-LPH plasma levels were
elevated in a large proportion of chronic schizophrenics,confirming
our previous studies in smaller groups of patients(5-7).The disso-
ciation of the values of the two peptides in half of the cases would
suggest either an alteration in the enzymatic breakdown of the
parental molecule proopiomelanocortin (POMC) in the pituitary,or
a different clearence rate or distribution volume of the two peptides. Alternatively, β-EP and β-LPH may partly derive from different
secretory pools.In animals, the anterior pituitary secretes POMC,
cleaved in ACTH ,β-LPH and β-EP in equimolar amounts.This secretion
is under the inhibitory control of noradrenaline(NE)(31).The intermediate pituitary secretes β-EP and alpha-MSH,with an inhibitory

154

control by dopamine(DA)(32). In man the existence of an anatomical
intermediate pituitary has not been demonstrated,but an intermedia-
te-like type of secretion has been suggested in some physiological
and pathological situations(33-36).Therefore, the dissociated levels
of β-EP and β-LPH observed in our schizohrenics may indicate that
two different pools of secretion are operative in these patients,
with the increased activity of one possibly being related to NE
deficiency resulting in parallel increases of basal β-EP and β-LPH
levels,while the other possibly being related to DA deficiency,with
increases in β-EP levels only.

The blunted responses of both β-EP and β-LPH to insulin stimulation and dexamethasone inhibition might be indicative of the presence of NE deficiency in a subgroup of schizophrenics(37) .
On the contrary,the lack of constant effects of neuroleptic therapy
on β-EP levels seems to exclude that the DA regulation of the opioid is impaired. Moreover,it would be difficult to propose the
existence of a DA deficiency in schizophrenia since a increased DA
tonus has been considered to be responsible for the disease(38).
NE deficiency has also never been clearly demonstrated. However,
the reduced GH responses to clonidine and apomorphine stimulation
observed in some chronic patients (39-40) may suggest that in these
subjects the whole catecholaminergic system , including NE and DA,
is somehow less active.

These preliminary data need further confirmation in order to
offer a more solid pathogenetic significance. Nonetheless, they
do indicate that the study of the peripheral opioid secretion can
reveal the existence of impairments of their regulatory mechanisms
in the hypothalamus and,possibly,in higher brain centers similar
to those observed in other neuroendocrine axes,which could be at
the basis of the mental disorder.

REFERENCES

1. Emrich,HM,Cording ,C,Pirèe, S, Kolling, A ,Muller ,HS ,Zerssen,
von , D and Herz , A (1979).Action of naloxone in different types
of psychoses .In: Usdin, E, Bunney, WE Jr. and Kline, NS (eds)
"Endorphins in mental health research".p.452(London:MacMillan Press)
2 . Ross, M, Berger, PA and Goldstein, A (1979).Plasma β-endorphin
immunoreactivity in schizophrenia. Science, 205, 1163
3. Berger, PA (1981).Clinical studies on the role of endorphins
in schizophrenia. III World Congr.Biol. Psychiat., Stockholm,Sweden,
Abstracts
4. Bianco, F, Castro, R and Sanchez, C (1981). Endorphins and
schizophrenia. III World Congr. Biol. Psychiat., Stockholm, Sweden,

Abstracts

5. Brambilla, F, Genazzani, AR ,Facchinetti, F, Parrini, D, Petraglia, F, Sacchetti, E, Scarone, S, Guastalla, A and D'Antona, N
(1981).Beta-endorphin and beta-lipotropin plasma levels in chronic
schizophrenia,primary affective disorders and secondary affective
disorders. Psychoneuroendocrinology, 6, 321

6. Brambilla, F, Facchinetti, F,Petraglia, F, Parrini, D,Vanzulli,
L and Genazzani, AR (1983).Endogenous opioid levels in chronic
schizophrenia.In: Endroczi ,E, De Wied, D, Angelucci, L and Scapagnini, U (eds)" Integrative neurohumoral mechanisms".p.515 (Amster-
dam:Elsevier Biomedical Press)

7. Brambilla, F, Facchinetti ,F and Genazzani ,AR(1984).Endogenous
opioid peptides in schizophrenia and affective disorders.In: Shah,
NS and Donald, AG(eds)"Psychoneuroendocrine dysfunction" .p.309.
(New York: Plenum Pub.Co.)

8. Pickar, D, Vartanian, F, Bunney, WE Jr, Mayer, HP, Gastpar, MT,
Prakash, G, Sethi, BB, Lideman, R, Belayev, B, Tsutsulkovskaja,
MVA, Jungkunz, G,Nedopil, N,Verhoeven, W and Van Praag,H (1982).
Short-term naloxone administration in schizophrenic and manic pa-
tients. Arch.Gen.Psychiat., 39, 313

9. Naber, D, Albus, M,Muller, F,Burke, H,Welter, D,Ackeneil, M
and Hippius,H (1982) .β-endorphin,cortisol and prolactin in serum
of schizophrenic patients during long-term neuroleptic treatment
and after withdrawal:relationship to psychopathology. Pharmacopsy-
chiatria, 2, 43

10. Weizman, R, Weizman, A, Tyano, S, Szekely, G,Weissman, BA and
 Sarne,Y (1984).Humoral endorphins blood levels in autistic,schizo-
phrenic and healthy subjects. Psychopharmacol. ,82 ,368

11. Terenius, L, Wahlström, A, Lindström, L and Widerlöw, E (1976)
Increased CSF levels of endorphins in chronic psychosis. Neurosci.
Letters, 3 ,157

12. Lindström, LH, Widerlöw, E, Gunne, LM, Wahlström, A and Terenius, L (1978).Endorphins in human cerebrospinal fluid:clinical
correlations to some psychotic states. Acta Psychiat.Scand.,57,153

13. Dupont ,A, Villeneuve, A and Bouchard, JP (1978).Rapid inacti-
vation of enkephalin-like material by CSF in chronic schizophrenia.
Lancet, ii ,1170

14. Höllt, V, Emrich, HM, Muller ,R and Falbusch, R (1978).
β-endorphin-like immunoreactivity (β-EI) in human plasma and cerebrospinal fluid.In : Van Ree, JM and Terenius,L (eds)"Characteristics and function of opioids " . p. 279 (Amsterdam :Elsevier/North
Holland)

15. Gunne, LM, Lindström, L and Widerlöw, E (1979) .Possible role
of endorphins in schizophrenia and other psychiatric disorders.In:
Usdin, E,Bunney, WE Jr and Kline, NS (eds)" Endorphins in mental
health research". p.547 (London :McMillan Press Ltd)
16. Domschke, W, Dickschas ,A and Mitznegg, P(1979).C.S.F. beta-
endorphin in schizophrenia. Lancet, i,1024
17. Nakao, K, Oki, S, Tanaka, I, Horii, K, Nakai, Y, Furui, T,
Fukushima, M, Kuwayama, A,Kageyama, N and Imura, H (1980).Immunore-
active beta-endorphin and adrenocorticotropin in human cerebrospinal fluid. J.Clin.Invest. ,66, 1383
18. Meurice ,E, Godon, JP and Mathieu ,F (1980). Les essais de
traitement de la schizophrenie par dialyse. Acta Psychiat. Belg.,80,
436
19. Rimon, R, Terenius, L and Kampman, R (1980).Cerebrospinal
fluid endorphins in schizophrenia. Acta Psychiat. Scand.,61, 395
20. Wahlström, A, Rehfeld, JF, Lindström, L and Terenius,L (1981).
Endorphins in cerebrospinal fluid of schizophrenics-present status.
III World Congr. Biol. Psychiat. ,Stockholm,Sweden, Abstracts
21. Van Kammen, DP,Waters, RN, Gold, P, Sternberg, D, Robertson,G,
Ganten, D, Pickar, D, Naber, D, Ballenger ,JC, Kaye, WH,Post, RM,
and Bunney, WE Jr (1981).Spinal fluid vasopressin,angiotensin I
and II,beta-endorphin and opioid activity in schizophrenia:a preliminary evaluation. In: Perris, C, Struwe, G and Jansson, B (eds)
"Biological Psychiatry" . p. 339 (Amsterdam: Elsevier/Nort Holland)
22. Buchsbaum ,MS, Davis, G, Naber, D, Pickar, D, Ballenger,J,
Waters, R, Goodwin, FK,Van Kammen, P,Post, R and Bunney, WE Jr
(1981)Pain appreciation,somatosensory evoked potentials and endorphins in normal and patients with schizophrenia. III World Congr.
Biol. Psychiat. ,Stockholm,Sweden ,Abstracts
23. Catlin, DH, Gerner, RH and Gorelick, DA (1981).Beta-endorphin:
behavioral effects of single and multiple infusions-measurement of
CSF levels. III World Congr.Biol.Psychiat., Stockholm,Sweden,
Abstracts
24. Pickar, D, Cohen, MR, Naber, D and Cohen, RM (1982).Clinical
studies of the endogenous opioid system.Biol.Psychiat., 17, 1243
25. Gerner, RH and Sharp ,B (1982) .CSF β-endorphin-immunoreacti-
vity in normal,schizophrenic,depressed,manic and anorexic subjects.
Brain Res., 237, 244
26. Lo Wen. ,HL and. Ho, WKK (1983). Cerebrospinal fluid Met^5-enke-
falin. level in schizophrenics during treatment with naloxone.
Europ.J.Pharmacol. ,92 ,77

27. Oliver, C, Mical, RS and Porter ,JC (1977) .Hypothalamic-pitui-
tary vasculature.Evidence for retrograde blood flow in the pituita-

ry stalk .Endocrinology,101, 598

28. Kastin, AJ, Olson, RD, Schally, AV and Coy, DH (1979). CNS effects of peripherally administered brain peptides.Life Sci.,25,401

29. Phillips, MJ (1984) Angiotensin and drinking-a model for the study of peptide action in the brain. In: Nemeroff, CB and Dunn,AJ (eds) " Peptides,hormones and behavior " .p. 423 (Lancaster-England: Spectrum Publ.Inc)

30. Facchinetti ,F and Genazzani,AR (1979).Simultaneous radioimmunoassay for beta-lipotropin and beta-endorphin in human plasma.In: Albertini,A, Da Prada, M and Peskar,BA (eds)"Radioimmunoassay of drugs and hormones in cardiovascular medicine". p.347(Amsterdam: Elsevier/North Holland)

31.Vermes,I,Mulder,GH,Smelik,PG and Tilders,FJH(1980) .Differential control of β-endorphin/β-lipotropin secretion from anterior and intermediate lobes on the rat pituitary gland in vitro.Life Sci., 27, 1761

32. McLoughlin,L,Tomlin, S, Grossman, A, Lytras, N,Schally,AV,Coy,D and Besser, GM (1984) .CRF-41 stimulates the release of β-lipotropin and β-endorphin in normal human subjects.Neuroendocrinology ,38 ,284

33. Wilkers, MM, Watkins, WB, Stuart, RD and Yen, SSC (1980).Localization and quantitation of β-endorphin in the human brain and pituitary.Neuroendocrinology, 30 ,113

34. Lamberts ,SWJ ,De Lange, SA and Stefanko ,SZ (1982).Adrenocorticotropin-secreting pituitary adenomas originate from the anterior or the intermediate lobe in Cushing disease:differences in the regulation of hormone secretion. J.Clin.Endocrinol.Metab. ,54 ,284

35. Facchinetti ,F ,Giovanni ,C ,Barletta ,C ,Petraglia ,F ,Buzzetti, R ,Burla ,F ,Lazzari ,R ,Genazzani ,AR and Scavo ,P (1986). Hyperendorphinemia in obesity and relationship to affective state. Psychology and Behav. ,36 ,937

36. Genazzani, AR ,Facchinetti, F ,Petraglia ,F, Pintor ,C, and Corda ,R ,(1986). Hyperendorphinemia in obese children and adolescents.J.Clin.Endocrinol.Metab. ,62 ,36

37. Müller, EE, Nisticò ,G and Scapagnini ,U.(1977)"Neurotransmitters and anterior pituitary function" (New York :Academic Press)

38. Carlsson, A (1978) .Mechanism of action of neuroleptic drugs. In:Lipton,MA,Di Mascio,A and Killam,KF(eds)"Psychopharmacology:a generation of progress" p.1057 (New York: Raven Press)

39. Rotrosen, J, Angrist, BM,Gershon, S, Sachar, EJ and Halpern,FS (1978).Neuroendocrine assestment of dopaminergic activity in schizophrenia.In : Costa, E and Gessa, GL (eds) " Advances in biochemical pharmacology" .p.649 (New York :Raven Press)

40. Muller-Spahn,F ,Ackeneil, M, Albus, M, Botschev ,C ,Naber,D and Welter, D (1986) Neuroendocrine effects of clonidine in chronic schizophrenic patients under long-term neuroleptic therapy and after drug withdrawal:relations to psychopathology.Psychopharmacol.,88,190

17

Outcome after 10 years of apomorphine-tested psychotic patients

A. Bocchetta, M. Del Zompo, G. Martis, S. Mulas and G.U. Corsini

INTRODUCTION

The antipsychotic efficacy of apomorphine and other dopamine agonists is still controversial matter. At the beginning of this
century Douglas [1] noted that apomorphine has hypnotic effects.
Bleuler [2] later described apomorphine as a "chemical restraint
of a special type" that calmed patients when given in an emetic
dose. In 1945 Feldman et al [3] noted the sedative effects of apomorphine in schizophrenics. In 1977 our group [4] observed sedative, hypnotic and antipsychotic effects of subemetic doses of
apomorphine in manic, schizoaffective, paranoid, hebephrenic and
catatonic patients. Since then several Authors [5-10] have reported improvement in schizophrenic patients after low-dose dopamine
agonists, whereas others [11-15] have not observed any benefit
(Table 1). The rationale of such studies is that, while conventional antipsychotic agents act through a blockade of postsynaptic
dopamine receptors, low-dose dopamine agonists may preferentially
stimulate presynaptic receptors, resulting in inhibition of excess
dopaminergic activity to which psychoses may be in part due [16-
17] . The existence of dopamine autoreceptors, whose stimulation
can inhibit the activity of the dopamine neuron on which they lie,
is suggested by behavioural, neurochemical, and electrophysiological studies [18-20].

In our preliminary report [4] we described sudden and striking reduction of the psychotic symptomatology lasting 20 to 50
minutes in 24 of 65 different unmedicated actively ill psychotic
patients, who had been injected intramuscularly with 10 μg/kg apomorphine 5 minutes earlier. Subsequently we selected for a retrospective study [21] eighteen patients from the original trial who
had a well documented clear-cut response to apomorphine, either
positive or negative. Diagnoses were carefully revised applying
the Research Diagnostic Criteria (RDC) [22], which had, at that

Table 1. Clinical studies of dopamine-agonists in schizophrenia

Source and year	Dose (mg)	N. improved
Apomorphine		
Corsini et al. 1977	1 i.m.	15/40
Smith et al. 1977 * §	1.5 - 6 s.c.	3/4
Tamminga et al. 1978 * §	3 s.c.	9/18
Hollister et al. 1980 * §	10 - 30 / day os	0/15
Meltzer 1980	0.75 s.c.	0/22
Ferrier 1982 §	0.75 s.c.	0/30
Cutler et al. 1982 * §	0.005 / kg s.c.	2/5
Levy et al. 1984 §	0.75 s.c.	0/25
Bromocriptine		
Tamminga et al. 1979 §	5 - 35 os	0/6
Meltzer et al. 1983	0.5 - 6 / day os	5/9
Cutler et al. 1984 * §	2 os	10/11
N-propylnorapomorphine		
Tamminga et al. 1986 §	5 - 40 os	5/18

* concurrent neuroleptic treatment
§ double-blind study

time recently appeared. Unexpectedly, and partially in contrast with the previous report, we found a marked difference of apomorphine responsiveness among RDC groups. Diagnosis in most of the responders (78%) was of schizoaffective disorder of manic type, whereas 89% of non-responders were diagnosed as schizophrenics. We also found that apomorphine-responders, as a group, appeared to have better outcome ratings, and a good response to lithium.

In the present study we extend the observation of the same patients to a ten-year-field follow-up, applying now a multiaxial evaluation according to the third edition of the Diagnostic and Statistical Manual of Mental Disorders (DSM III) [23].

SUBJECTS AND METHODS

The subjects in this study were 18 patients who had been tested with apomorphine (10 µg/kg i.m.) 8 to 10 years ago during a psychotic episode that required hospitalisation. They had been selected from the original 65 patients of a previous report [4] for a subsequent retrospective study [21] on the basis of their clearcut response to apomorphine, either positive (more than 75% improvement of Brief Psychiatric Rating Scale [24] score) or negative (less than 20% improvement). The subjects were traced, contacted, and personally interviewed by two psychiatrists blind to the previous diagnosis, and to the apomorphine-response. Outcome status ratings in four areas (marital, residential, occupational, psychiatric) were assessed according to Tsuang et al [25]. Consensus diagnosis according to DSM III [23] was reached using the interview material, information from relatives, and medical records. The level of adaptive functioning during the past year was evaluated through axis V of DSM III.

RESULTS

The application of DSM III resulted in rather less homogeneous diagnoses compared to RDC (Table 2). From the original two main RDC groups of the previous study (schizoaffective disorder and schizophrenia) the new classification yielded four groups (affective disorder with mood-incongruent psychotic features, schizoaffective, schizophreniform, and schizophrenic disorders). Nevertheless DSM III provided interesting information about the adaptive functioning of the patients during the past year. Significantly higher levels were found in apomorphine-responsive subjects (Table 3). The same trend was observed in the outcome status ratings in all the investigated areas but the marital one (Table 4).

Table 5 and 6 show the diagnoses of apomorphine-responders and non-responders, respectively. The group of RDC schizoaffect-

161

Table 2. Diagnoses according to different criteria

RDC diagnosis	n.	DSM III diagnosis	n.
Schizoaffective disorder	8	Affective disorder with mood-incongruent psychotic features	5
Schizophrenia	10	Schizoaffective disorder	1
		Schizophreniform disorder	5
		Schizophrenia	7

Table 3. Adaptive functioning during the past year

Level according to Axis V of DSM III	Responders	Non-responders
Good - fair	78%	33%
Poor - very poor - grossly impaired	22%	67%

Table 4. Outcome status according to Tsuang et al.

Area	Rating	Responders	Non-responders
Occupational	Good	78%	44%
	Poor	11%	33%
Residential	Good	89%	56%
	Poor	11%	22%
Psychiatric	Good	44%	22%
	Poor	22%	33%

Note : Intermediate ratings and marital status not shown

ives had the highest percentage of positive responses (87%). On the contrary most of schizophrenics, however diagnosed, failed to improve.

Table 5. Comparison between diagnoses of responders

	RDC	DSM III
1)	Schizoaffective, manic type	Bipolar disorder, mixed, mood-incongruent psychotic features
2)	Schizoaffective, manic type	Bipolar disorder, manic, mood-incongruent psychotic features
3)	Schizoaffective, manic type	Bipolar disorder, manic, mood-incongruent psychotic features
4)	Schizoaffective, manic type	Bipolar disorder, manic, mood-incongruent psychotic features
5)	Schizoaffective, manic type	Schizoaffective disorder
6)	Schizoaffective, manic type	Schizophreniform disorder
7)	Schizoaffective, manic type	Schizophreniform disorder
8)	Schizophrenia, paranoid	Schizophrenia, paranoid
9)	Schizophrenia, disorganized	Schizophrenia, disorganized

DISCUSSION

The main purpose of this study was to clarify if patients who had significantly improved after low-dose apomorphine during an acute psychotic episode might share common features, and if apomorphine and other dopamine agonists may, at least, represent a tool to predict outcome or response to conventional treatments.

The application of different diagnostic criteria has provided interesting information even though it may have added confusion to some important points. Thus, while RDC proved to be more useful than DSM III in the classification of single acute episodes sus-

Table 6. Comparison between diagnoses of non-responders

	RDC	DSM III
1)	Schizoaffective, depressive type	Major depression, recurrent, mood-incongruent psychotic features
2)	Schizophrenia, paranoid	Schizophreniform disorder
3)	Schizophrenia, paranoid	Schizophreniform disorder
4)	Schizophrenia, disorganized	Schizophreniform disorder
5)	Schizophrenia, paranoid	Schizophrenia, paranoid
6)	Schizophrenia, paranoid	Schizophrenia, paranoid
7)	Schizophrenia, paranoid	Schizophrenia, paranoid
8)	Schizophrenia, disorganized	Schizophrenia, disorganized
9)	Schizophrenia, disorganized	Schizophrenia, disorganized

ceptible of improvement with apomorphine, DSM III with its multi-axial evaluation gave better insight into the good outcome of apo-morphine-responsive patients.

Our data suggest that low-dose apomorphine may act predomi-nantly on acute psychotic episodes with manic features rather than on chronic schizophrenics or psychotic depressed patients. However no definite conclusion can be drawn, also considering that the debate on the antipsychotic effect of dopamine agonists has re-ceived contrasting contributions in the past years. The discrepan-cies may be explained in view of several methodological aspects implied in such studies. The various Authors may have included different subgroups of patients, due to the use of different diag-nostic criteria. Moreover, blind conditions cannot be easily main-tained because of the multiple side effects of dopamine agonists. Another source of discrepancy may be the rather wide range of doses used, or the concurrent administration of neuroleptics. It might be even supposed that two different apomorphine-induced phe-nomena were actually observed. The improvement of mental conditions in chronic schizophrenics under neuroleptic medication might be similar to that observed after levodopa, especially with respect to negative symptoms [26,27]. Finally, Tamminga et al [15] suggest that the antipsychotic action of dopamine agonists may be predominant in a subgroup of subjects with neuroleptic-responsive symptoms, not in neuroleptic-nonresponders.

In our hands apomorphine seems to act primarily during episodes with manic features, as evidenced by RDC but regardless of the principal DSM III diagnosis. This opinion is corroborated by the observation of a patient who had responded to apomorphine during a schizoaffective manic episode, but failed to gain any beneficial effect from the drug during a subsequent schizoaffective depressive episode. The good response to lithium as well as the better outcome ratings of apomorphine-responders is in agreement with such an opinion. Furthermore, Tesarova [28] has reported experimental depression caused by apomorphine in volunteers, Post et al [29] have noted antimanic properties of apomorphine in a psychotic patient, whereas Meltzer [30] has described mood changes in apomorphine-treated schizophrenics. Ferrier et al [13], even though not observing any specific therapeutic effect, have demonstrated a reduction in anxiety in acute schizophrenics.

Whether apomorphine can specifically reduce psychotic symptoms in schizophrenics or simply induces sedation with nonspecific reduction in all behaviours is difficult to ascertain. In our responsive patients, marked formal thought disorders with multiple delusions and hallucinations completely subsided in few minutes, and a conceptual re-organization was observed. Unfortunately such a "miracle" lasts less than an hour, and tolerance occurs when dopamine agonists are administered repeatedly [15] , limiting at present the potential clinical usefulness of this approach. Nevertheless we can confirm here that apomorphine may be used, as a test, to predict the response to conventional treatments (lithium, neuroleptics), and the short and long-term outcome in psychotic patients.

REFERENCES

1.   Douglas, CJ (1900). Apomorphine as a hypnotic. N Y Med J, 71, 376
2.   Bleuler, E (1911). "Dementia praecox, or the group of schizophrenias". p.486. (New York: International Universities Press)
3.   Feldman, F, Susselmann, S and Barrera, SE (1945). A note on apomorphine as a sedative. Am J Psychiatry, 102, 403

4. Corsini, GU, Del Zompo, M, Manconi, S, Cianchetti, C, Mangoni, A and Gessa, GL (1977). Sedative, hypnotic, and antipsychotic effects of low doses of apomorphine in man. In: Costa, E and Gessa, GL (eds.) "Advances in Biochemical Psychopharmacology – Vol.16". p.645. (New York: Raven Press)

5. Smith, RC, Tamminga, C and Davis, JM (1977). Effect of apomorphine on schizophrenic symptoms. J Neural Transm, 40, 171

6. Tamminga, CA, Schaffer, MH, Smith, RC and Davis, JM (1978). Schizophrenic symptoms improve with apomorphine. Science, 200, 567

7. Cutler, NR, Jeste, DV, Karoum, F and Wyatt, RJ (1982). Low-dose apomorphine reduces serum homovanillic acid concentrations in schizophrenic patients. Life Sci, 30, 753

8. Meltzer, HY, Kolakowska, T, Robertson, A and Tricou, BJ (1983). Effect of low-dose bromocriptine in treatment of psychosis: the dopamine autoreceptor-stimulation strategy. Psychopharmacology, 81, 37

9. Cutler, NR, Jeste, DV, Kaufmann, CA, Karoum, F, Schran, HF and Wyatt, RJ (1984). Low dose bromocriptine: a study of acute effects in chronic medicated schizophrenics. Prog Neuro-Psychopharmacol Biol Psychiat, 8, 277

10. Tamminga, CA, Gotts, MD, Thaker, GK, Alphs, LD and Foster, NL (1986). Dopamine agonist treatment of schizophrenia with N-propyl-norapomorphine. Arch Gen Psychiatry, 43, 398

11. Hollister, LE, Davis, KL and Berger, PA (1980). Apomorphine in schizophrenia. Commun Psychopharmacol, 4, 277

12. Meltzer, HY (1980). Relevance of dopamine autoreceptors for psychiatry: preclinical and clinical studies. Schizophr Bull, 6, 456

13. Ferrier, IN (1982). Clinical and hormonal effects of apomorphine in acute and chronic schizophrenia. Br J Psychiatry, 140, 204

14. Levy, MI, Davis, BM, Mohs, RC, Kendler, KS, Mathé, AA, Trigos, G, Horvath, TB and Davis, KL (1984). Apomorphine and schizophrenia. Arch Gen Psychiatry, 41, 520

15. Tamminga, CA and Schaffer, MH (1979). Treatment of schizophrenia with ergot derivatives. Psychopharmacology, 66, 239

16. Carlsson, A (1978). Does dopamine have a role in schizophrenia? Biol Psychiatry, 13, 3

17. Skirboll, LR, Grace, AA and Bunney, BS (1979). Dopamine auto- and post-synaptic receptors: electrophysiological evidence for differential sensitivity to dopamine agonists. Science, 206, 80

18. Aghajanian, GK and Bunney, BS (1976). Dopamine autoreceptors: pharmacological characterization by microiontophoretic single cell recording studies. Naunyn Scmiedebergs Arch Pharmacol, 297, 1

19. Roth, RH (1979). Dopamine autoreceptors: pharmacology, func-

tion and comparison with post-synaptic dopamine receptors. Commun Psychopharmacol, 3, 429

20. Corsini, GU, Del Zompo, M, Manconi, S, Piccardi, MP, Onali, PL, Mangoni, A and Gessa, GL (1978). Evidence for dopamine receptors in the human brain mediating sedation and sleep. Life Sci, 20, 1613

21. Del Zompo, M, Pitzalis, GF, Bernardi, F, Bocchetta, A and Corsini, GU (1981). Antipsychotic effect of apomorphine: a retrospective study. In: Corsini, GU and Gessa, GL (eds.) "Apomorphine and Other Dopaminomimetics. Vol.2. Clinical Pharmacology". p.65. (New York: Raven Press)

22. Spitzer, RL, Endicott, J and Robins, E (1978). Research Diagnostic Criteria (RDC) for a Selected Group of Functional Disorders, ed.3. (New York: State Psychiatric Institute)

23. American Psychiatric Association (1980). Diagnostic and Statistical Manual of Mental Disorders, ed.3 (DSM III)

24. Overall, JE and Gorham, DR (1962). The Brief Psychiatric Rating Scale. Psychol Rep, 10, 799

25. Coryell, W and Tsuang, MT (1986). Outcome after 40 years in DSM-III schizophreniform disorder. Arch Gen Psychiatry, 43, 324

26. Gerlach, J and Lühdorf, K (1975). The effect of L-dopa on young patients with simple schizophrenia, treated with neuroleptic drugs. A double-blind cross-over trial with Madopar and placebo. Psychopharmacologia, 44, 105

27. Ogura, C, Kishimoto, A and Nakao, T (1976). Clinical effect of L-dopa on schizophrenia. Curr Ther Res, 20, 308

28. Tesarova, O (1972). Experimental depression caused by apomorphine and phenoharmane. Pharmakopsychiatrie, 5, 13

29. Post, RM, Gerner, RH, Carman, JS and Bunney Jr, WE (1976). Effects of low doses of a dopamine-receptor stimulator in mania. Lancet, i, 203

30. Meltzer, HY (1982). Dopamine autoreceptor stimulation: clinical significance. Pharmacol Biochem Behav, 17, 1

167

18
Noradrenaline, psychotic decompensation, and drug response in schizophrenia

D.P. van Kammen, T.C. Neylan,
W.B. van Kammen, J.L. Peters,
J. Rosen and M. Linnoila

INTRODUCTION

Recent biological investigation in schizophrenia has revealed a role for noradrenaline (NA). Disturbed noradrenergic activity is among the most replicated biological findings in schizophrenia. Increased NA levels have been found in CSF, autopsy brains, and plasma. For review see Hornykiewicz [1] and van Kammen and Antelman [2]. However, not all investigators have demonstrated increased NA indices of activity. CSF NA has been found to be highly variable from LP to LP, in contrast to CSF MHPG, HVA and 5HIAA [3]. Alpha-2 receptor function has been found to be both increased [4-6] and decreased [7-9]. This confusing variability of results left some investigators to conclude that the noradrenergic system was not disturbed in schizophrenia [10].

Besides being coincidental, disturbed NA activity can be important to schizophrenia in three ways:

1. Trait Markers.
    Cross-sectional studies have produced mainly conflicting if not paradoxical results. Typically, these findings have been explained by a heterogeneous or multifactorial etiology of schizophrenia [11].

2. Illness Modifiers.
    Noradrenaline could influence course and expression of the illness without causing the illness. CSF DBH and platelet monoamine oxidase may be examples of such modifiers.

3. State Dependent Markers.
    Inconsistent findings in noradrenergic activity and schizophrenia may be

explained by state dependent changes. Significant variance is accounted for by episodic changes in clinical state reflecting coinciding changes in the underlying biochemistry.

STAGES OF PSYCHOTIC DECOMPENSATION

Recently, new attention has been focused upon prodromal phenomena and the process of psychotic decompensation. Several authors have discussed the different stages and symptomatology of the psychotic decompensation process [12-14]. Psychotic decompensation appears to unfold in a characteristic way in many patients. Docherty *et al.* [14] have identified these stages: (1) overextension - the subject first begins to feel stressed, (2) restrictive consciousness or withdrawal phase in which negative symptoms appear, (3) disinhibition, (4) psychotic disintegration, and (5) psychotic resolution.

Prodromal symptoms become more observable [12] in the disinhibition of prepsychotic phase. Hyperarousal and loss of sleep are the hallmarks of the pre-, or early psychotic stages. The hyperarousal can be explained either as secondary to the psychotic symptoms or as a necessary condition for psychotic decompensation. Overarousal may be present also during the overextension stage. Underarousal may be present during the restrictive consciousness stage. Since patients remain for an extended period of time in any one stage, this may well explain the large literature on over-, and underarousal observed in schizophrenic patients as well as the two types of schizophrenia concept [15]. Recent evidence indicates that some negative symptoms respond to neuroleptic treatment, that with increased duration of illness negative symptoms become treatment resistant, and that these negative symptoms may be present prior to the illness and coincide with impairments in functioning. This suggests to us that the two types [15] may refer to state dependent phenomena [16].

THE NORADRENALINE-DOPAMINE HYPOTHESIS OF PSYCHOTIC DECOMPENSATION AND ANTIPSYCHOTIC DRUG RESPONSE

What leads to this episodic dysregulation? King *et al.* [17] have explored it in the intrinsic dopamine regulatory mechanism. In this paper we will make the case that a disturbance in the noradrenergic system may trigger this dysregulation of the dopamine system. We postulate that during the prodromal stages, e.g. during biochemical and clinical instability, excess NA release increases arousal and psychotic behavior, decreases sleep, and throws the impaired but compensated dopamine regulation in disarray. When NA release normalizes, sleep duration returns to normal, but the psychotic episode runs its course.

CSF NA, SLEEP, AND PSYCHOSIS

Kemali *et al.* [18] reported that CSF NA levels correlated significantly with daytime EEG indices of arousal in schizophrenic patients. CSF NA has been found to be significantly higher in the patients who slept less hours the night before the LP (r = -0.44, N = 53, p = 0.0008) [2]. This relationship and that CSF NA levels correlated with psychosis [3] may explain the variability of CSF NA levels of several drug-free LPs within the same patients. We noted that CSF NA levels declined with pimozide treatment parallel to the clinical improvement (r = 0.71, p < 0.02) [7]. It is of interest that disturbed sleep is frequently among the first symptoms to respond to neuroleptic treatment. In the figure, we show daily psychosis ratings and hours of sleep (at night) plotted together in three day running means. During the prodromal phase, prior to reaching relapse criteria, sleep decreases while psychosis increases. After relapse is established, sleep "normalizes", while psychosis continues to increase. The exact role of NA in sleep and wakefulness is uncertain, but it may act as a modulator [19]. Our data suggest that interpretation of sleep EEG data in drug-free schizophrenic patients will be problematic unless prodromal symptoms and state dependent effects are controlled for.

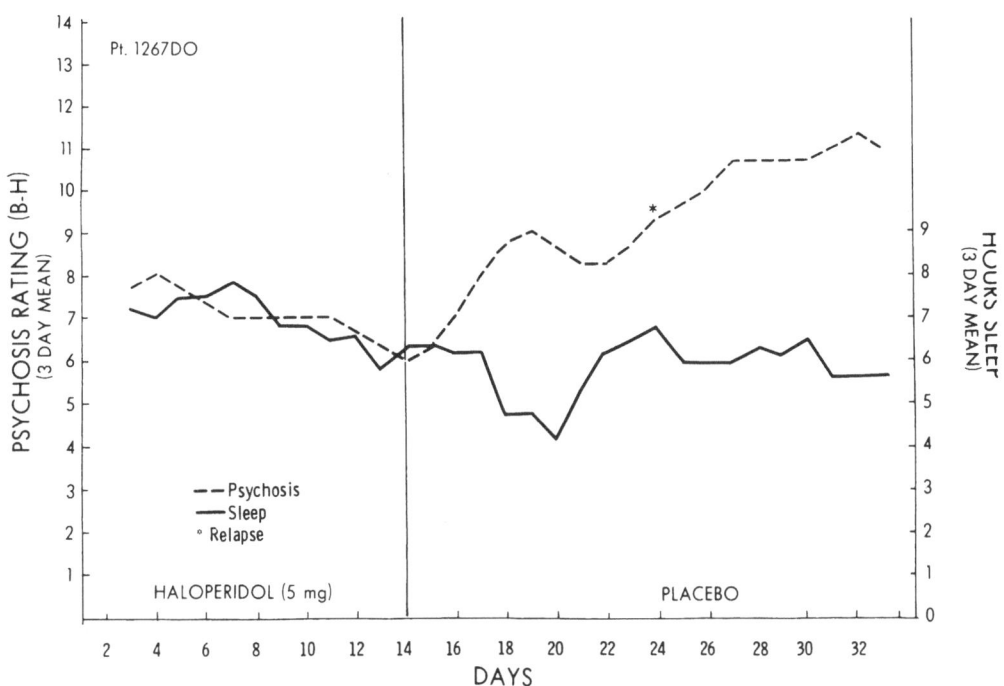

Figure 1 Relationship between sleep and psychotic exacerbation.

CSF MHPG, SLEEP, AND PSYCHOSIS

We have found in the past a relationship between CSF MHPG and psychosis (r = 0.66, N = 15, p < 0.01), but not in all groups of drug-free schizophrenic patients (van Kammen *et al.*, unpublished data). To follow up on this inconsistent finding, we studied 28 physically healthy male schizophrenic patients (DSM-III), who received an LP during haloperidol treatment and again after drug withdrawal within days of meeting relapse criteria (N = 12) or after six weeksif relapse had not been observed (N = 16). During the drug-free period CSF MHPG, the hours of sleep the night before, and the psychosis ratings of the day before the LP correlated significantly with each other, but not with CSF HVA. During haloperidol treatment such relationships were not observed (Table 1). Surprisingly, the connection was only significant in the patients who subsequently relapsed after drug withdrawal (r = 0.65, p = 0.01), and not in the haloperidol treated non-relapsers (r = 0.23, p = NS), even though CSF MHPG and the hours of sleep in the two groups on haloperidol were identical [20].

Table 1 Relationships between sleep, psychosis and CSF MHPG and HVA

A. Drug-free patients

	PSYCHOSIS			CSF MHPG			CSF HVA		
	r	N	p	r	N	p	r	N	p
Sleep	-0.65	27	0.0001	-0.55	27	0.001	-0.03	27	NS
Psychosis				0.49	28	0.004	-0.15	28	NS
CSF MHPG							0.24	28	NS

B. Haloperidol-treated patients

	r	N	p	r	N	p	r	N	p
Sleep	-0.25	27	NS	0.41	27	0.016	0.19	26	NS
Psychosis				0.02	27	NS	-0.20	26	NS
CSF MHPG							0.20	27	NS

CSF AND PLASMA MHPG AND DRUG RESPONSE

We have reported previously that CSF MHPG correlated significantly with the change in psychosis following an infusion of 20 mg d-amphetamine in drug-free schizophrenia patients (r = 0.58, N = 23, p < 0.002) [21]. Patients with high or low MHPG values worsened respectively. We have in the past interpreted that the behavioral change induced by d-amphetamine indicated that the patient is in an unstable condition. A similar relationship for CSF HVA was not observed. We found also that patients who improved with d-amphetamine were more likely to improve with pimozide treatment. Recently, plasma HVA and MHPG has gained some interest as being under state dependent control. Bowers et al. [22] observed that both were higher in patients who showed a rapid response to neuroleptics. Plasma MHPG may reflect CNS levels, whereas plasma HVA may not.

We found CSF NA to correlate significantly with DA ($r = 0.64$, $N = 34$, $p < 0.0001$) and dopamine sulfate ($r = 0.35$, $N = 35$, $p < 0.05$) [2]. Since NA and DA do interact, a dysregulation of the NA system may trigger the psychotic episode that only through changes in the DA system can be aborted. We propose that increased release of NA may be necessary but not sufficient for behavioral change, be it worsening or improvement, drug-induced or "spontaneous". One could speculate that the increased NA release may result from a faulty feedback regulation because of impaired alpha-2 receptor function. It is unknown whether alpha-2 receptor binding is state dependent. Further studies will be needed to clarify this issue.

We reviewed the evidence that increased NA release, whether it is expressed in CSF levels of NA or MHPG, accompanies increased psychosis and decreased sleep. This is particularly present during the early phases or stages of psychotic decompensation but not during haloperidol treatment. We propose that the increased NA release, either causing or being caused by alpha-2 receptor dysfunction, may trigger a psychotic episode which requires an impaired or overactive DA system and which runs its course unless it is treated with neuroleptic drugs. These elevated NA levels may facilitate faster neuroleptic response. Although NA and DA are separate systems in the brain, they share influence on many behaviors such as attention and arousal. We conclude that both NA and DA play a role in psychotic decompensation.

REFERENCES

1. Hornykiewicz, O (1982). Brain catecholamines in schizophrenia - a good case for noradrenalines. *Nature*, **299**, 484
2. van Kammen, DP and Antelman, S (1984). Impaired noradrenergic transmission in schizophrenia? A minireview. *Life Sci*, **34**, 1403
3. Linnoila, M, Ninan, PT, Scheinin, M, Waters, RN, Chang, W, Bartko, J and van Kammen, DP (1983). Reliability of norepinephrine and major monoamine metabolites measurements in CSF of schizophrenic patients. *Arch Gen Psychiatry*, **40**, 1290
4. Matussek, N, Ackenheil, M, Hippius, H, Mueller, F, Schroder, HT, Shultes, H and Wasilewski, B (1980). Effect of clonidine on growth hormone release in psychiatric patients and controls. *Psychiatry Res*, **2**, 25
5. Kafka, MS, Siever, LJ, Nurnberger, JI, Uhde, TW, Targum, S, Cooper, DMJ, van Kammen, DP and Tokola, (1985). Platelet alpha-adrenergic receptor function in affective disorders and schizophrenia. *Psychopharmacol Bull*, **21**, 599
6. Pandey, GN, Garver, DL, Tamminga, CA, Ericksen, S, Ali, SI and Davis, JM (1977). Postsynaptic supersensitivity in schizophrenia. *Am J Psychiatry*, **134**, 518
7. Sternberg, DE, van Kammen, DP, Lake, CR, Marder, SR and Bunney, WE, Jr. (1981). The effect of pimozide on CSF norepinephrine in schizophrenia. *Am J Psychiatry*, **138**, 1045

8.   Ko, GN, Unnerstall, JR, Kuhar, MJ, Wyatt, RJ and Kleinman, JE (1986). Alpha-2 adrenergic agonist binding in schizophrenic brains. *Psychopharmacol Bull*, **22**, 1011

9.   Mller-Spahn, F, Ackenheil, M, Albus, M, Botschev, C, Naber, D and Welter, D (1986). Neuroendocrine effects of clonidine in chronic schizophrenic patients under long-term neuroleptic therapy and after drug withdrawal: relations to psychopathology. Psychopharmacology, 88, 190

10.   Gattaz, WF, Riederer, P, Reynolds, GP, Gattaz, D and Beckman, H (1983). Dopamine and noradrenaline in cerebrospinal fluid of schizophrenic patients. *Psychiatry Res*, **8**, 243

11.   Lake, CR, Sternberg, D, van Kammen, DP, Ballenger, JC, Ziegler, MG, Post, RM, Kopin, IJ and Bunney, WE, Jr. (1980). Schizophrenia: elevated cerebrospinal fluid norepinephrine. *Science*, **207**, 331

12.   Herz, MI, Szymanski, HV and Simon, JC (1982). Intermittent medication for stable schizophrenic outpatients: an alternative to maintenance medication. *Am J Psychiatry*, **139**, 918

13.   Szymanski, HV, Simon, JC and Gutterman, N (1983). Recovery from schizophrenia psychosis. *Am J Psychiatry*, **140**, 381

14.   Docherty, JP, van Kammen, DP, Siris, SG and Marder, SR (1978). Stages of onset of schizophrenic psychosis. *Am J Psychiatry*, **135**, 420

15.   Crow, TJ (1980). Molecular pathology of schizophrenia: more than one disease process? *Br Med J*, **280**, 66

16.   van Kammen, DP, Rosen, J, Peters, J, Fields, R and van Kammen, WB (1985). Are there state dependent markers in schizophrenia? Psychopharmacol Bull, 21, 497

17.   King, R, Barchas, JD and Huberman, BA (1984). Chaotic behavior in dopamine neurodynamics. *Proc Natl Acad Sci USA*, **81**, 1244

18.   Kemali, D, Maj, M, Iorio, G, Marciano, F, Nolfe, G, Galderisi, S and Salvati, A (1985). Relationship between CSF noradrenaline levels, C-EEG indicators of activation and psychosis ratings in drug-free schizophrenic patients. *Acta Psychiatr. Scand.*, **71**, 19

19.   Monti, JM (1983). Catecholamines and the sleep-wake cycle. II. REM sleep. *Life Sci*, **32**, 1401

20.   van Kanmmen, DP, van Kammen, WB, Peters, JL, Rosen, J, Slawsky, RC, Neylan, TC, Linnoila, M (1986). CSF MHPG, sleep and psychosis in schizophrenia. *Clin Neuropharm*, **9**, suppl. 4, 575

21.   van Kammen, DP, Bunney, WE, Jr., Docherty, JP, Marder, SR, Eberg, MH, Rosenblatt, JE and Rayner, JN (1982). d-Amphetamine-induced heterogeneous changes in psychotic behavior in schizophreniz. *Am J Psychiatry*, **139**, 991

22.   Bowers, MB, Jr., Swigar, ME, Jatlow, PI, Hoffman, F and Giocoechea, N (1986). Early neuroleptic response in psychotic men and women: correlation with plasma HVA and MHPG. *Comp Psychiatry*, **27**, 181

19
The biochemical basis for the antipsychotic effects of neuroleptics

D. Pickar, A. Breier, O.M. Wolkowitz and A.R. Doran

The introduction of chlorpromazine into clinical practice in the 1950's represented a landmark not only for psychiatry but for all of medicine [1]. In addition to providing new treatments for schizophrenia and psychosis in general, antipsychotic drugs, referred to generically as neuroleptics, provided a major lead for understanding the pathophysiology of the most serious of mental illnesses [2].

The first evidence linking the mechanism of action of neuroleptic drugs to CNS dopamine systems was derived from the pioneering work of Carlsson and Lindqvist [3,4]. It was observed by these investigators that neuroleptic administration increased the accumulation of dopamine metabolites in rat brain. The hypothesis that this metabolite increase occurs in response to neuroleptic-induced post-synaptic receptor blockade of CNS dopamine neurons is supported by numerous subsequent studies and now stands as a fundamental construct of neuropharmacology [5].

The development of techniques by which receptors are labelled and receptor-ligand interactions defined advanced the link between neuroleptic action and "blockade" of CNS dopamine neurotransmission. It has been shown that the affinity of individual antipsychotic drugs for the butyrophenone labelled, non-adenyl cyclase D_2 receptor is closely correlated with their potency as antipsychotics [6,7]. These findings are unique in their demonstration of an association between in vitro pharmacology and complex clinical actions. Thus, chlorpromazine whose antipsychotic doses may range from 300 mg to 1.0 gm/day or greater has less affinity for the D_2 dopamine receptor than does fluphenazine whose antipsychotic doses are in the range of 10-50 mg/day [1].

TIME-DEPENDENT PHARMACOLOGY OF NEUROLEPTICS

Perhaps the greatest limitation to the receptor blockade model for neuroleptic action is the fact their antipsychotic effects of neuroleptic drugs require weeks or months to be optimal whereas receptor blockade occurs rapidly after the initiation of neuroleptic therapy. The fact that rapid neuroleptization, i.e., progressive increases in neuroleptic dosage during the first 24-48 hours of treatment as a strategy to maximize receptor blockade [8], provides little advantage over standard neuroleptic administration schedules in hastening the antipsychotic response [1], underscores the short-comings of the receptor blockade model of antipsychotic mechanism.

In recent years, preclinical studies have begun to examine slow to develop effects of neuroleptic drugs on CNS dopamine systems. Evidence from both electrophysiologic [9] and biochemical [5] studies suggest that dopamine neurons respond differently when acute versus chronic administration schedules are compared. Whereas neuroleptic drugs characteristically produce rapid and marked increases in dopamine neuronal firing in nigro-striatal and mesolimbic systems in response to receptor blockade, prolonged neuroleptic administration (weeks) produces markedly decreased rates of individual neuronal firing with complete loss of spontaneous activity in some neurons [9-12]. Time-dependent changes in biochemical indices of dopamine "turnover" have also been observed in various animal brain regions [5,13,14]. Consistent with electrophysiologic data, acute neuroleptic administration greatly increases dopamine turnover as reflected by the increased accumulation of brain levels of homovanillic acid (HVA); prolonged neuroleptic administration, however, is associated with reductions in dopamine turnover to, but not below, pretreatment baseline. Thus, both electrophysiologic and biochemical data suggest neuroleptic-induced time-dependent changes in dopamine neuronal function. These phenomena may have important implications for the clinical action of neuroleptics.

CLINICAL STUDIES OF PLASMA HVA IN SCHIZOPHRENIA

Consistent with the discrepancy between the time course of antipsychotic effects and of receptor blockade, clinical markers of neuroleptic-induced dopamine receptor blockade, e.g., neuroleptic-induced elevation in even circulating levels of prolactin, the emergence of extrapyramidal symptoms or circulating levels of dopamine receptor blocking potency as determined by radioreceptor assay, have been poor predictors of clinical response [15]. In order to examine the hypothesis that slow to develop changes in dopamine function are involved in the mechanism of neuroleptic antipsychotic action, we have applied the strategy of longitudinal assessment of plasma levels of HVA, a major dopamine metabolite, to schizophrenic patients during the course of neuroleptic treatment [16-18]. Some caution is needed in interpreting these levels with regard to CNS dopamine activity, however, since levels of HVA which circulate in plasma are comprised of both peripheral and central nervous system contributions. Peripheral sources of dopamine are thought to be largely derived from the peripheral sympathetic nervous system as well as from the adrenal medulla and the kidney. Although the exact proportion of central and peripheral contributions to plasma levels of HVA is unknown, animal experimentation has suggested that circulating levels of HVA parallel alterations in CNS dopamine system turnover produced by pharmacologic stimuli [19,20]. It has been estimated that 40-60% of circulating levels of HVA owe their original source to the CNS [21].

We have observed that the administration of the antipsychotic neuroleptic, fluphenazine, to schizophrenic patients is associated with slow to develop decreases in levels of plasma HVA [16,17]. In these experiments, total free plasma HVA was assessed by high pressure liquid chromatography with electrochemical detection [22] in plasma samples taken by venipuncture between 7:00am and 8:00am, three times per week, following overnight fast and during a period of restricted activity.

Figure 1 shows mean weekly levels of plasma HVA from 16 DSM-III diagnosed schizophrenic patients beginning from a placebo baseline (mean ± SEM days free from all active medications: 34 ± 6 days) and extending through 5 subsequent weeks of fluphenazine treatment (mean ± SEM mg/day: 29.5 ± 3.6). As seen in Figure 1, significant fluphenazine-induced decreases were observed only after the third week of treatment. The importance of these time-delayed alterations in dopamine function are suggested by the high correlations between nurses' global psychosis ratings (made by staff who were "blind" to medication status) and levels of plasma HVA when all points of collections were considered (3 per week) (Figure 2). These data suggest that for this group of patients, symptom reduction, as reflected by ratings of global psychosis, parallels decreasing levels of plasma HVA. In addition to the similar time course of plasma HVA and psychosis reduction for the overall patient group, we have found that in individual patients fluphenazine-induced reduction in mean weekly levels of plasma HVA are correlated with reduction in psychosis and negative symptoms ratings [17,18]. These findings suggest that the antipsychotic effects of fluphenazine may be related to reductions in dopamine turnover which occur only after continued drug administration. In this regard, these clinical data bear resembalance to neuroleptic-induced changes in dopamine function observed after prolonged administration in limbic and striatal regions of rodent and non-human primate brain [23]. Whereas receptor blockade may ultimately be the initiating event of neuroleptic action, feedback reductions in dopamine turnover may be more closely related to the antipsychotic mechanism.

MESOLIMBIC AND MESOCORTICAL DOPAMINE SYSTEMS

An important recent neuroscience development has been the delineation of a discrete group of CNS dopamine neurons which originate in the midbrain and innervate the neocortex. This so-called mesocortical dopamine system is anatomically associated with the mesolimbic dopamine neurons. In contrast to mesolimbic neurons, however, mesocortical dopamine neurons display a series of unique characteristics. In addition to maintaining a higher rate of spontaneous firing at rest, mesocortical neurons show less response to both dopamine agonists and antagonists; moreover, they do not show the characteristic time-dependent changes in response to chronic neuroleptic administration seen in mesolimbic and nigro-striatal neurons [24]. The clinical relevance of the mesocortical

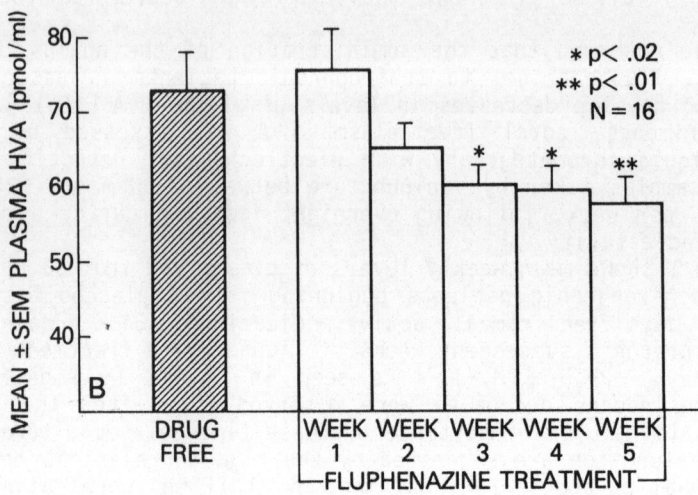

FIGURE 1   Time-dependent decreases in mean weekly levels of plasma HVA during fluphenazine treatment of 16 DSM-III diagnosed schizophrenic patients (F = 6.59, p<0.001) (17)

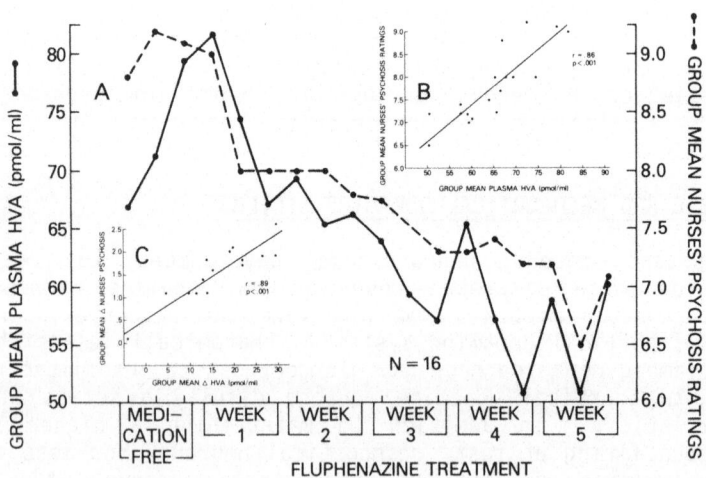

FIGURE 2   Levels of plasma HVA and nurses' psychosis ratings and their correlations for all data points (17)

dopamine system is suggested since it has been shown to modulate limbic dopaminergic activity [25], to be involved in mediating frontal cortical behaviors [26] and to be uniquely stress-responsive [27]. The implications of the differential neuroleptic response patterns of mesolimbic and mesocortical dopamine neurons is not fully understood. Our clinical data indicating neuroleptic-induced time-dependent changes in dopamine turnover suggest the possibility that the mesolimbic response may be fundamental to neuroleptic clinical action. Enchancement of mesocortical response by pharmacologic strategies such as the administration of alprazolam as an adjunct to neuroleptic treatment [28] may prove to be useful in augmenting neuroleptic response.

REFERENCES

1. Baldessarini, RJ (1980). Drugs and the treatment of psychiatric disorders. In: Gilman, BH, Goodman, LS and Gilman, A (eds.), "The Pharmacologic Basis of Therapeutics." (New York: McMillan)
2. Meltzer, HY and Stahl, SM (1976). The dopamine hypothesis of schizophrenia: a review. Schizophr Bull, 2, 19-76
3. Carlsson, A and Lindqvist, M (1963). Effect of chlorpromazine or haloperidol on formation of 3-methoxytyramine and normetanephrine in mouse brain. Acta Pharmacol Toxicol, 20, 140-144
4. Carlsson, A (1978). Antipsychotic drugs, neurotransmitters and schizophrenia. Am J Psychiatry, 135, 164-173
5. Roth, RH (1983). Neuroleptics: functional chemistry. In: Coyle, JT and Enna, SJ (eds.), "Neuroleptics: Neurochemical, Behavioral and Clinical Perspectives." (New York: Raven Press)
6. Creese, I, Burt, DR and Snyder, SH (1976). Dopamine receptor binding predicts clinical and pharmacological potencies of anti-schizophrenic drugs. Science, 192, 481-483
7. Seeman, P, Lee, T, Chau-Wong, M and Wong, K (1976). Anti-psychotic drug doses and neuroleptic/dopamine receptors. Nature, 261, 717-719
8. Donlon, PT and Tupin, JP (1975). Rapid "digitalization" of decompensated schizophrenic patients with antipsychotic agents. Am J Psychiatry, 132, 1023-1026
9. Bunney, BS (1984). Antipsychotic drug effects on the electrical activity of dopaminergic neurons. Trends in Neuroscience Research, 212-215
10. Bunney, BS and Grace, AA (1978). Acute and chronic haloperidol treatment: comparison of effects on nigral dopaminergic cell activity. Life Sci, 23, 1715-1728
11. Chiodo, LA and Bunney, BS (1983). Typical and atypical neuroleptics: differential effects of chronic administration on the activity of A_9 and A_{10} midbrain dopaminergic neurons. J Neurosci, 3, 1607-1619
12. White, JF and Wang, RY (1983). Differential effects of classical and atypical antipsychotic drugs on A_9 and A_{10} dopamine neurons. Science, 221, 1054-1056
13. Bacopoulos, NG, Buntos, G, Redmond, DE, et al (1978). Regional sensitivity of primate brain dopaminergic neurons to haloperidol: alterations following chronic treatment. Brain Res, 157, 396-401

14. Bacopoulos, NG, Redmond, DE, Baulu, J and Roth, RH (1980). Chronic haloperidol or fluphenazine: effects on dopamine metabolism in brain, cerebrospinal fluid and plasma of cercopithecus aethiops (Vervet monkey). J Pharmacol Exp Ther, 212, 1-5

15. Meltzer, HY, Kane, JM and Kalakowska, T (1983). Plasma levels of neuroleptics, prolactin levels, and clinical response, In: Coyle, JT and Enna, SJ (eds.), "Neuroleptics, Neurochemical, Behavioral and Clinical Perspectives." pp. 255-279. (New York: Raven Press)

16. Pickar, D, Labarca, R, Linnoila, M, Roy, A, Hommer, D, Everett, D and Paul, SM (1984). Neuroleptic-induced decrease in plasma homovanillic acid and antipsychotic activity in schizophrenic patients. Science, 225, 954-957

17. Pickar, D, Labarca, R, Doran, AR, Wolkowitz, OM, Roy, A, Breier, A, Linnoila, M and Paul, SM (1986). Longitudinal measurement of plasma homovanillic acid levels in schizophrenic patients: correlation with psychosis and response to neuroleptic treatment. Arch Gen Psychiatry, 43, 669-676

18. Pickar, D, Wolkowitz, O, Labarca R, Doran AR, Breier A, Paul SM (1987). Biochemical alterations produced by neuroleptics in man: studies of plasma homovanillic acid in schizophrenic patients. In: Dahl, SG, Gram, LF, Paul, SM and Potter, WZ (eds.), "Clinical Pharmacology in Psychiatry" (Vol 3). pp. 248-254. (New York: Springer-Verlag)

19. Bacopoulos, NG, Hattox, SE and Roth, RH (1979). 3,4-dihydroxy-phenylacetic acid and homovanillic acid in rat plasma: possible indicators of central dopaminergic activity. Eur J Pharmacol, 56, 225-236

20. Kendler, KS, Heninger, GR and Roth, RH (1982). Influence of dopamine agonists on plasma and brain levels of homovanillic acid. Life Sci, 30, 2063-2069

21. Sternberg, DE, Heninger, GR and Roth, RH (1983). Plasma homovanillic acid as an index of brain dopamine metabolism: enhancement with debrisoquin. Life Sci, 32, 2447-2452

22. Chang, WH, Sheinin, M, Burns, RS, et al (1983). Rapid and simple dertermination of homovanillic acid in plasma using high performance liquid chromatography with electrochemical detection. Acta Pharmacol Toxicol, 53, 275-279

23. Pickar, D (1986). Neuroleptics, dopamine, and schizophrenia. Psychiatr Clin North Am, 9(1), 35-48

24. Bannon, MJ and Roth, RH (1983). Pharmacology of mesocortical dopamine neurons. Pharmacologic Reviews, 35, 53-68

25. Pycock, CJ, Kerwin, RW and Carter, CJ (1980). Effect of lesion of cortical dopamine terminals on subcortical dopamine in rats. Nature, 286, 74-77

26. Milner, B and Petrides, M (1984). Behavioral effects of frontal lobe lesions in man. Trends in Neuroscience, 7, 403-407

27. Tam, SY and Roth, RH (1985). Selective increase in dopamine metabolism in the prefrontal cortex by the anxiogenic β-carboline FG-7142. Biochem Pharmacol 34, 1595-1598

28. Wolkowitz OM, Pickar D, Doran AR, Breier A, Tarell J, Paul SM (1986). Combination alprazolam-neuroleptic treatment of the positive and negative symptoms of schizophrenia. Am J Psychiatry 143(1), 85-87

20
Schizophrenia: new windows
C.L. Cazzullo

In recent years, schizophrenia has undoubtedly attracted the scientific world's more active interest.

This fact is to be associated with a different type of approach to such a widespread problem which is increasingly referred to as "the schizophrenic group", as E. Bleuler pointed out in 1911.

I would like to stress that in the recent past the awareness of the interaction of various factors in the pathogenesis of the disease has led to an integration of many tests by means of which they can be recognized. Moreover, it seems that cross-analysis has permitted us to obtain useful information all the more valuable because it can be internally validated.

The concept of an integrative psychiatry able to employ cross-correlations has been significantly represented at this Meeting.

The identification of risk factors appears to be as important as the so called markers of outcome of schizophrenic illness. The former are related to the subject's vulnerability to the disease and the latter pertain to its progressive course over time.

Along these lines, the first issue to be considered is that of epidemiology within the framework of demographic and biological variables.

This entails consideration of some fundamental individual variables such as birthdate, sex, age at onset and HLA typing. We have tried to correlate these with the symptomatological heterogeneities of the schizophrenic disorders.

We have been able to confirm, at least as a trend, the birth seasonality data from other European countries in a large sample of Italian patients (1). We found an excess of schizophrenic births in November and December as compared to the live general population born in the same period (1930-1952), without significant differences between disorganized or mixed and paranoid types, thus supporting the hypothesis of the involvement of environmental factors in the etiology of the disorder.

A "viral" hypothesis relating to some groups of schizophrenic patients (and so clearly presented at this Meeting) deserves further attention if we consider the analogies with some viral or suspected slow-viral diseases such as multiple sclerosis. The intermittent course of the disease in so many patients, the exposure period between 12 and 15 years of age and the T lymphocyte behavioral pattern suggest the possibility of a revival of a slow virus in patients with a defined immunological asset. Consequently, I would suggest devoting rapid and particular attention to the group of HTLV I-IV viruses as postulated by some advanced research centers in the U.S.A., for example, the Vistar Institute, in Philadelphia.

As far as age at onset is concerned, we have focused our attention on the patients' sex, diagnostic/clinical subtype and HLA typing (2). Two sequential studies showed that all these variables play a significant part in the variability of the age at onset of schizophrenic disorders, but without any reciprocal interaction.

We found an earlier age at onset in the disorganized and undifferentiated subtypes (DSM III), that is, earlier in the non-paranoid (severe) as opposed to the paranoid (mild) forms.

The significant effect of HLA type on age at onset is represented by an earlier onset of the disorder in HLA-A1 and /or CRAG A1 positive subjects than in A1 and /or CRAG A1 negative ones.

The effect of the sex variable showed an earlier distribution of age at onset for males than for females. The average distibution in females is delayed by 4-5 years in comparison with that in males. This fact, independent of biological, hormonal or environmental origins, implies that females have greater chances of being affected by a less severe form of illness, while males are at a higher risk of a more global destructuration of the personality.

The role of the sex variable is not limited to the presence or absence of the disease, but also involves the specificity of the symptomatological profile and, therefore, the clinical aspect of the disease at its onset. The later disease onset in females implies important symptomatological differences.

In male patients dissociative elements and their most important behavioral consequences tend to prevail, whereas alterations of the emotional-instinctual sphere and of affectivity occur more frequently in females. This kind of psychopatological analysis reminds the basic symptoms described by Bleuler, incoherence, loosening of associations, inappropriate affect.

According to more recent clinical conceptualizations, part of these symptoms may represent alogia. In fact, in one of our studies we observed a group of schizophrenic patients among whom alogia was prominent and precocious, mainly found in male patients with early onset and a poor prognosis.

Among the various indices under genetic control the more promising seem to be: a) the HLA antigens, and b) the smooth pursuit eye movements (SPEM).

a) - Starting from our first study in this field (3) numerous associations between schizophrenia and class I (A, B, C) HLA antigens have been found. Association between HLA A1 and the hebephrenic type of schizophrenia and between HLA A9 and paranoid forms, were subsequently found (4).

These associations might suggest, inter alia, involvement of the immune system in schizophrenia. Recent research, however, has shown that most of the immune diseases appear to be primarily associated with class II antigens of the Major Histocompatibility Complex (MHC), especially with DR antigens. Consequently, analysis of DR antigens could be a more valid tool for further investigation of the autoimmune hypothesis of schizophrenia. For this reason, we have recently analized the HLA-DR antigens of 74 schizophrenic patients (5).

We found that the DR3 antigen was significantly overrepresented in patients ($X^2 = 6.1$, $p < .02$) as compared with healthy controls. Consequently, the risk of schizophrenia in a DR3-positive individual was twice as high as that in one who lacked this antigen (relative risk = 2.01) without any specific relationship to one or another subtype of the disease.

This association was particularly pronounced in patients with both birth complications ($p < .025$) and suicidal behavior ($p < .05$).

At this point, the autoimmune characteristics of schizophrenia need further clarification as far as cross-correlations are concerned. This should be undertaken with some other parameter like VBR, positive or negative family history, sex and age of onset.
b) - As far as SPEM are concerned, a result of one cross correlation appears to be particularly worthy of mention here.

When compared to normal controls, a group of Italian schizophrenic patients had SPEM measures that were significantly abnormal (6). Our results are consistent with those reported by others and confirm the existence of a higher prevalence of bad tracking performances in schizophrenics than in normal controls. The concept that SPEM dysfunctions in schizophrenia should be considered a "trait" rather than a "state" marker, and that they may be related to the genetic systems controlling susceptibility to schizophrenia is widely accepted, and has been confirmed by our preliminary analysis.

Turning to the markers of outcome of schizophrenic disorders, we know that a great deal of attention was given to brain structural abnormalities. These were detected by pneumoencephalography back in the sixties (7).

The second section of this Meeting is largerly devoted to this topic. The most relevant results reported are almost entirely consistent with those of our group (8,9).

A fourth of our schizophrenic population presented definite ventricular enlargement. Cross-correlation analysis showed , in this subgroup of patients, a significant increase of the HLA-B12 antigen, a scarse positive family history for schizophrenia, normal alfa band on computerized EEG, some degree of cognitive impairment, poor response to neuroleptic treatment, and a poor prognosis. Curious enough was the observation of an increase of VBR among patients born in the winter months (10).

Intensive research in Experimental Allergic Encephalomyelitis (EAE), brain injury, hydrocephalus, epilepsy have shown that ventricular enlargement is a non-specific brain lesion. In fact, this kind of pathology is primarily confined to the ependyma and the basal structure of the wall of the ventricles. Any type of stimulus produces an inflammatory reaction of which edema is the most relevant feature.

For this reason one must inquire as to what really is the prime mover of the ventricular pathology, that secondarily induces a periventricular or a perimeningeal brain atrophy.

A neurofunctional parameter such as computerized EEG is related to some fundamental functions and can be useful for a deeper analysis of such symptoms as illogical thinking and /or flatness of affect.

Past research has shown the particular balance of hemispheric functioning in schizophrenia. This is essentially an overactivation of the left hemisphere or, better, a dysregulation of neurofunctional pattern. The clinical heterogeneity of schizophrenic patients and differences in their psychopathological histories, however, do not permit statements about the universality of this pattern. It seems that diagnostic subgroups, clinical symptomatology and drug regimen might be critical in determining neurofunctional assessments in schizophrenia.

Some new laboratory data can substantiate this hypothesis.

On one hand, we analyzed the distribution of hemispheric-related errors on a neuropsychological test, the Quality Extinction Test (QET), as a function of the disease chronicity (11).

When compared with normal controls, schizophrenic patients showed a higher incidence of left extinguishing responses to the QET. Moreover, chronic patients had more left-side impairment than patients with subchronic forms of schizophrenia. These results suggest therefore a strong influence of the disease course on hemispheric impairment.

On the other hand, we studied the influence of a series of clinical and epidemiological characteristics on the EEG frequency bands in schizophrenic and normal subjects (12). Diagnosis alone did not satisfactorily explain the power characteristics. On the contrary sex, age and drug regimen had the greatest effects on them.

A new line of research is offered by recording of the Movement Related Brain Macropotentials (MRBMs). This provides a tentative neurophysiological model of adult cognitive processes both in normal and pathological subjects.

When normal subjects were compared with chronic schizophrenics, the Bereitschaftspotential (BP) amplitude of the latter was reduced and the Skilled Performance Positivity (SPP) was absent (13).

The biochemistry of schizophrenia is obviously a very important field because it can be directly related to the effects of treatment better than any other. In this regard, more attention is now being paid to the use of non-neuroleptic drugs such as hormones, anticonvulsant compounds, cortisone, etc., for treatment of schizophrenia.

Finally, I suggest that the contribution of the basic sciences, genetics, biochemistry, neuropathology are as important as the psychopathological characteristics related to age, sex and disease course. We will continue to discuss of "familial" and "sporadic" schizophrenia, but we still are not sure about the extent of the heterogeneity in the former, nor, in particular, in the latter.

Integrative psychiatry therefore is not a presumption, but a promising reality.

REFERENCES

1.   Cazzullo, CL, Zacchetti, L, Morabito, A, Piacentini, D, Provenza, M, Ronchi, P and Bellodi, L (1985). Seasonal nature of births of schizophrenic patients in the Italian population. In:Cazzullo, CL and Invernizzi, G (eds) "Schizophrenia: An Integrative View". p.293. (London: John Libbey)
2.   Bellodi, L, Smeraldi, E, Orsini, A and Cazzullo, CL (1985). Effect of HLA type on age at onset in schizophrenic disorders. Neuropsychobiol., 14, 62
3.   Cazzullo, CL, Smeraldi, E and Penati, G (1974). The leukocyte antigenic system HLA as a possible genetic marker of schizophrenia. Brit. J. Psychiatry, 125, 25
4.   Smeraldi, E, Bellodi, L and Cazzullo, CL (1976). Further studies on the major histocompatibility complex as a genetic marker for schizophrenia. Biol. Psychiatry, 11, 655
5.   Cazzullo, CL, Vita, A, Sacchetti, E, Illeni, MT, Ciussani, S and Invernizzi, G (1986). HLA-DR antigens in schizophrenia. Integrative Psychiat., 3, 199
6.   Smeraldi, E, Gambini, O, Bellodi, L, Sacchetti, E, Vita, A, Di Rosa, M, Macciardi, F and Cazzullo, CL (in press). Combined measure of smooth pursuit eye movements (SPEM) and ventricular brain ratio (VBR) in patients with schizophrenic disorders. Psychiatry Res.
7.   Cazzullo, CL (1963). Biological and clinical studies on schizophrenia related to pharmacological treatment. Rec. Adv. Biol. Psychiat., 5, 114
8.   Vita, A, Sacchetti, E, Calzeroni, A, Invernizzi, G and Cazzullo, CL (1986). Computed tomographic findings in psychiatric disorders. In: Reisner, T, et al. (eds) "Advances in Neuroimaging". p.243. (Wien: Verlag der Wiener Medizinischen Akademie)
9.   Sacchetti, E, Vita, A, Calzeroni, A, Invernizzi, G and Cazzullo CL (this volume). Neuromorphological correlates of schizophrenic disorders: focus on cerebral ventricular enlargement
10.   Sacchetti, E, Vita, A, Battaglia, M, Calzeroni, A, Conte, G, Invernizzi, G and Cazzullo, CL (this volume). Season of birth and cerebral ventricular enlargement in schizophrenia
11.   Gambini, O, Cazzullo, CL and Scarone, S (1986). Interpretation of abnormal responses to the Quality Extinction Test in schizophrenia. J. Neurol. Neurosurg. Psychiat., 49, 997
12.   Scarone, S, Pugnetti, L, Cattaneo, AM, Biserni, P, Capuano, P, Colombo, C, Cattaneo, R, Gambini, O and Cazzullo CL (1986). Electrophsysiological indexes of hemispheric abnormalities in schizophrenia: clinical and epidemiological correlates. In: Shagass, C et al. (eds) "Biological Psychiatry 1985". p.1054. (Amsterdam: Elsevier Science Publ.)
13.   Chiarenza, GA, Papakostopoulos, D, Dini, M and Cazzullo, CL (1985). Neurophysiological correlates of psychomotor activity in chronic schizophrenics. Electroencephalogr. and Clin. Neurophysiol., 61, 218

Index

189